Lecture Notes in Physics

For information about Vols. 1–110, please contact your bookseller or Springer-Verlag.

Vol. 111: H.-J. Schmidt, Axiomatic Characterization of Physical Geometry. V, 163 pages. 1979.

Vol. 112: Imaging Processes and Coherence in Physics. Proceedings, 1979. Edited by M. Schlenker et al. XIX, 577 pages. 1980.

Vol. 113: Recent Advances in the Quantum Theory of Polymers. Proceedings 1979. Edited by J.-M. André et al. V, 306 pages. 1980.

Vol. 114: Stellar Turbulence. Proceedings, 1979. Edited by D. F. Gray and J. L. Linsky. IX, 308 pages. 1980.

Vol. 115: Modern Trends in the Theory of Condensed Matter. Proceedings, 1979. Edited by A. Pekalski and J. A. Przystawa. IX, 597 pages. 1980.

Vol. 116: Mathematical Problems in Theoretical Physics. Proceedings, 1979. Edited by K. Osterwalder. VIII, 412 pages. 1980.

Vol. 117: Deep-Inelastic and Fusion Reactions with Heavy Ions. Proceedings, 1979. Edited by W. von Oertzen. XIII, 394 pages. 1980.

Vol. 118: Quantum Chromodynamics. Proceedings, 1979. Edited by J. L. Alonso and R. Tarrach. IX, 424 pages. 1980.

Vol. 119: Nuclear Spectroscopy. Proceedings, 1979. Edited by G. F. Bertsch and D. Kurath. VII, 250 pages. 1980.

Vol. 120: Nonlinear Evolution Equations and Dynamical Systems. Proceedings, 1979. Edited by M. Boiti, F. Pempinelli and G. Soliani. VI, 368 pages. 1980.

Vol. 121: F. W. Wiegel, Fluid Flow Through Porous Macromolecular Systems. V, 102 pages. 1980.

Vol. 122: New Developments in Semiconductor Physics. Proceedings, 1979. Edited by F. Beleznay et al. V, 276 pages. 1980.

Vol. 123: D. H. Mayer, The Ruelle-Araki Transfer Operator in Classical Statistical Mechanics. VIII, 154 pages. 1980.

Vol. 124: Gravitational Radiation, Collapsed Objects and Exact Solutions. Proceedings, 1979. Edited by C. Edwards. VI, 487 pages. 1980.

Vol. 125: Nonradial and Nonlinear Stellar Pulsation. Proceedings, 1980. Edited by H. A. Hill and W. A. Dziembowski. VIII, 497 pages. 1980.

Vol. 126: Complex Analysis, Microlocal Calculus and Relativistic Quantum Theory. Proceedings, 1979. Edited by D. Iagolnitzer. VIII, 502 pages. 1980.

Vol. 127: E. Sanchez-Palencia, Non-Homogeneous Media and Vibration Theory. IX, 398 pages. 1980.

Vol. 128: Neutron Spin Echo. Proceedings, 1979. Edited by F. Mezei. VI, 253 pages. 1980.

Vol. 129: Geometrical and Topological Methods in Gauge Theories. Proceedings, 1979. Edited by J. Harnad and S. Shnider. VIII, 155 pages. 1980.

Vol. 130: Mathematical Methods and Applications of Scattering Theory. Proceedings, 1979. Edited by J. A. DeSanto, A. W. Sáenz and W. W. Zachary. XIII, 331 pages. 1980.

Vol. 131: H. C. Fogedby, Theoretical Aspects of Mainly Low Dimensional Magnetic Systems. XI, 163 pages. 1980.

Vol. 132: Systems Far from Equilibrium. Proceedings, 1980. Edited by L. Garrido. XV, 403 pages. 1980.

Vol. 133: Narrow Gap Semiconductors Physics and Applications. Proceedings, 1979. Edited by W. Zawadzki. X, 572 pages. 1980.

Vol. 134: $\gamma\gamma$ Collisions. Proceedings, 1980. Edited by G. Cochard and P. Kessler. XIII, 400 pages. 1980.

Vol. 135: Group Theoretical Methods in Physics. Proceedings, 1980. Edited by K. B. Wolf. XXVI, 629 pages. 1980.

Vol. 136: The Role of Coherent Structures in Modelling Turbulence and Mixing. Proceedings 1980. Edited by J. Jimenez. XIII, 393 pages. 1981.

Vol. 137: From Collective States to Quarks in Nuclei. Edited by H. Arenhövel and A. M. Saruis. VII, 414 pages. 1981.

Vol. 138: The Many-Body Problem. Proceedings 1980. Edited by R. Guardiola and J. Ros. V, 374 pages. 1981.

Vol. 139: H. D. Doebner, Differential Geometric Methods in Mathematical Physics. Proceedings 1981. VII, 329 pages. 1981.

Vol. 140: P. Kramer, M. Saraceno, Geometry of the Time-Dependent Variational Principle in Quantum Mechanics. IV, 98 pages. 1981.

Vol. 141: Seventh International Conference on Numerical Methods in Fluid Dynamics. Proceedings. Edited by W. C. Reynolds and R. W. MacCormack. VIII, 485 pages. 1981.

Vol. 142: Recent Progress in Many-Body Theories. Proceedings. Edited by J. G. Zabolitzky, M. de Llano, M. Fortes and J. W. Clark. VIII, 479 pages. 1981.

Vol. 143: Present Status and Aims of Quantum Electrodynamics. Proceedings, 1980. Edited by G. Gräff, E. Klempt and G. Werth. VI, 302 pages. 1981.

Vol. 144: Topics in Nuclear Physics I. A Comprehensive Review of Recent Developments. Edited by T.T.S. Kuo and S.S.M. Wong. XX, 567 pages. 1981.

Vol. 145: Topics in Nuclear Physics II. A Comprehensive Review of Recent Developments. Proceedings 1980/81. Edited by T. T. S. Kuo and S. S. M. Wong. VIII, 571-1.082 pages. 1981.

Vol. 146: B. J. West, On the Simpler Aspects of Nonlinear Fluctuating. Deep Gravity Waves. VI, 341 pages. 1981.

Vol. 147: J. Messer, Temperature Dependent Thomas-Fermi Theory. IX, 131 pages. 1981.

Vol. 148: Advances in Fluid Mechanics. Proceedings, 1980. Edited by E. Krause. VII, 361 pages. 1981.

Vol. 149: Disordered Systems and Localization. Proceedings, 1981. Edited by C. Castellani, C. Castro, and L. Peliti. XII, 308 pages. 1981.

Vol. 150: N. Straumann, Allgemeine Relativitätstheorie und relativistische Astrophysik. VII, 418 Seiten. 1981.

Vol. 151: Integrable Quantum Field Theory. Proceedings, 1981. Edited by J. Hietarinta and C. Montonen. V, 251 pages. 1982.

Lecture Notes in Physics

Edited by H. Araki, Kyoto, J. Ehlers, München, K. Hepp, Zürich
R. Kippenhahn, München, H. A. Weidenmüller, Heidelberg
and J. Zittartz, Köln

187

Density Functional Theory

With contributions by

C. Amador, M. P. Das, R. A. Donnelly, J. L. Gázquez,
J. E. Harriman, J. Keller, M. Levy, J. P. Perdew,
A. Robledo, C. Varea, E. Zaremba

Edited by J. Keller and J. L. Gázquez

Springer-Verlag
Berlin Heidelberg GmbH 1983

Editors

Jaime Keller
José Luis Gázquez
D. E. Pg., Facultad de Quimica &
Facultad de Estudios Superiores, Cuautitlan
Universidad Nacional Autonoma de Mexico
04510, Mexico, D. F.

ISBN 978-3-540-12721-5 ISBN 978-3-540-38703-9 (eBook)
DOI 10.1007/978-3-540-38703-9

Originally published by Springer-Verlag Berlin Heidelberg New York in 1983
2153/3140-543210

PREFACE

Density functional theory has received a great deal of attention during the last few years both at a fundamental level and in applications to the determination of the electronic structure of atoms, molecules and solids.

The two major reviews written in the past: P. Gombas, "Die statistische Theorie des Atoms und ihre Anwendung" (Springer, Vienna 1949) and N.H. March, Adv. Phys. 6, 1 (1957), presented the Thomas-Fermi and the statistical exchange-correlation approximations before the theory was formalized through the Hohenberg-Kohn (1964) theorems.

A large number of papers have appeared in the last 20 years on the basic aspects and applications of density functional methods. The work on applications to matter is, for example, reviewed in J.C. Slater, "Quantum Theory of Molecules and Solids," Vol.4, (McGraw-Hill, New York 1974), and on chemical applications in A.S. Bamzai and B.M. Deb, Rev. Mod. Phys. 53, 95 (1981). Here the emphasis will be more on the basic aspects of density functional theory itself and on the practical philosophy of its use. The examples given are only meant to illustrate the results that can be expected or the practical motivations for the different approximations currently used.

The topics covered include:

1) Fundamental aspects of the Hohenberg-Kohn theorems.
2) Density matrices and reduced density matrices, fundamental aspects, properties and applications to density functional theory.
3) One-matrix energy functionals, fundamental aspects and properties.
4) Local density approximations from momentum space considerations with emphasis on the exchange energy functional.
5) Kinetic energy functionals of non-electron-gas type, derived from models of the electron first- and second-order density matrices.
6) Electron gas models within the Kohn-Sham formalism explaining fundamental aspects of the $X\alpha$ theory and developing new functionals for the exchange energy as well as local potentials for atoms from a non-electron gas type approach.
7) The self-interaction correction, its importance, its implications and applications to atomic properties.
8) Linear response theory in density functional theory, applications to atoms and solids.
9) Relativistic aspects of density functional theory.
10) Density functionals for nonuniform classical fluids.

The book is based on the "International Workshop on Electronic Density Functionals" that took place in México City, sponsored by Universidad Nacional Autónoma de México. It was decided to produce it in textbook form so that it could be used as an introduction to current research on this topic.

J. Keller

J.L. Gázquez

CONTENTS

INTRODUCTION

J. Keller

Density functional theory considers the particle density to be the fundamental variable to describe the state of a system in an external potential. The external potential itself may be static or time dependent and the system may be composed of particles obeying classical or quantum mechanics. In this book we will think of electronic densities as main example. The last chapter is however devoted to classical inhomogeneous systems. There are examples in the literature where the methods are applied to nuclear matter or to astrophysical objects.

Historically the density functional approach initiated with the idea that locally the behavior of a collection of particles, the electron cloud, could be represented and approximated by that of a free electron gas of the same density at that point. The Thomas-Fermi (TF) model [1,2] was in many aspects very succesful and showed the basic steps to obtain the density functional for the total energy: using standard quantum mechanics based on wave functions to obtain from a well defined model (and some extra assumptions, averages and practical simplifications guided by physical considerations) a direct relationship, functional, between the total energy and the density. The theory went even farther as it allowed the direct determination of a charge density for a given external potential.

The TF theory was also the first to exploit the idea of a <u>local</u> density functional when it considered a functional depending on the variables at only one point in space, the total kinetic energy being obtained by volume integration of a local kinetic energy density

$$E_\kappa = \int e_k(1)\, d\tau_1 \quad , \quad e_k(1) = e_k\{\rho(1)\} \qquad (1)$$

The electron-electron interaction (without exchange and correlation) was not a local functional , but simply a density functional as a double volume integral is required.

The TF theory, however, providing a differential equation for the self-consistent determination of the charge density without the intermediate use of wave functions, stands as a model for the study of a system without the use of wave equations.

The later inclusion, by Dirac (3) in 1930, of a local density functional for the electron-electron exchange energy and by Wigner (4) in 1934 of a local electron-electron correlation energy functional, made it clear that the density functional approach was feasible. But there were too many drawbacks: poor agreement with experimental total energies, difficulties with negative ions, electronic charge distributions too large at the nucleus and at large distances, no clear way to introduce a procedure to obtain a shell structure, molecules being unbound, etc. The method seemed destined to remain one of only qualitative value even with the von Weizsacker (5) corrections to the kinetic energy functional introducing terms in the local values of the gradients of the electronic density, and refinements (6) to it introduced even in recent years (see for example Huntington (7)).

There were three major reasons, however, for the density functional theory acquiring a relevant position in the last two decades: the formalization of the density functional theory itself, the many practical applications of the local density exchange-correlation approximation to reduce the many electron problem to a (selfconsistent) one electron effective potential and, finally, the better understanding of the electron gas with the many body techniques.

The formalization of the density functional theory started with the proof of the Hohenberg and Kohn (8) theorems in 1964 stating that the total energy of a many particle system in its ground state is a unique functional of its particle density and that there exists a variational principle for the energy functional. This made the density functional theory a viable alternative to wave function theories. More recently Levi (9) solved the n-representability problem and formally defined the universal functional of the density for the sum of the kinetic and potential energies. The formalism and theorems have also been generalized to the time-dependent external potential case (10-13). The method adquired a very useful formulation with the Kohn-Sham (14) series of equations where the charge density is obtained through a set of auxiliary functions (very often called themselves single particle wave functions)

$$\rho(\underline{r},\underline{r}') = \sum_i \Theta(\mu - \varepsilon_i)\ \psi_i(r)\psi_i^{+}(r') \qquad (2)$$

which obey, using the Hohenberg-Kohn variational principle, a set of self-consistent ("single particle") equations

$$H_{eff}\psi_i(r) = \varepsilon_i\psi_i(r) \qquad (3)$$

The chemical potential μ appears as a Lagrange multiplier, in the variational procedure, introducing the condition of total number of particles conservation:

$$\int n(\underline{r})\, d\tau = N; \qquad n(\underline{r}) = \sum_i (\Theta(\mu - \varepsilon_i)\psi_i^+(r)\psi_i(r) \qquad (4)$$

The chemical potential itself and its relation to the electronegativity χ

$$\mu = [dE/dN] = -\chi \qquad (5)$$

has also been studied.[15].

The derivation of the fundamental theorems is reviewed in chapters 2, 3 and 4 of this book.

The fact that the local density approximation allows the study of a many electron system by solving the auxiliary self - consistent single particle equations, (usually called the single particle Schrödinger equations) promoted a large number of applications in atomic, molecular and condensed matter physics and chemistry (see for example [16]). The large scale use of the local density exchange-correlation started with the band-structure calculations in the last 20 years (see for example [17,18] and references therein).

For magnetic materials or for the study of magnetic properties the different potentials for different spins approximation has been used, the method often takes the name local-spin-density and has been formalized by von Barth and Hedin [19].

The introduction of the multiple scattering techniques and of other methods to solve the auxiliary equations for the non periodic multicenter potential of a molecule [20] allowed a large number of studies which have been relevant to quantum chemistry and molecular physics in general.

The use of statistical exchange to reduce the Hartree-Fock problem to a one electron approximation is linked to a basic proposal by Slater (21) now known as the Xα method.

Another type of problem where density functional theory, in the sense at least of a local exchange-correlation one electron potential, has been fruitfully applied is the study of amorphous solids (22), of liquids (23), of impurities and dislocations (24), and even of crystalline solids (25) from a cluster method or real space approach.

A recent review can be found in the book of Avery and Dahl (26).

Other developments refer to the form the density functional should take. For the study of the kinetic energy term the electron gas was taken as a working model, as a result of that there is only one parameter in the theory: the free electron gas parameter $\rho^{1/3}$ as in the TF theory. Other terms have to be adimensional, for example the Weizsäcker term, have the dimensions of the gradient square $(\nabla\rho)^2/\rho^2$. For spherically symmetric systems (atoms) additional terms including $1/r$ or $\partial/\partial r$ can be considered (27). This is analysed in chapter 9 in this book.

Another set of physical constrains on the form of the density functional is given by the boundary conditions on the one- and two-particles density matrices (Kutzelnigg, del Re and Berthier (31)) for a system with a finite number n of electrons of a given spin. The pair correlation function for electrons of the same spin tends to $-1/n$ at large distances, this is negligible for an infinite system but important for atoms and molecules where n can be very small. Gopinathan, Whitehead and Bogdanovic (32) showed that this boundary condition could be used to derive the Z dependence of the α parameter in the Xα theory (33). Keller and Gázquez (34) have shown that the use of physically guided realistic forms for the pair correlation functions, with the finite number of electrons boundary condition built in, leads to local exchange-correlation-potentials for atoms and molecules which give total energies and exchange or correlation energies very close to the Hartree-Fock limit and to the experimental values. The corrections to the momentum at the Fermi level that result from this boundary condition and thus the dependence of the kinetic energies on the total number of electrons (35) (also of the Coulomb energies (36)) are reviewed in chapter 7.

Finally as the density functional formalism is not restricted to non-relativistic quantum theory classical (chapter 10) and quantum relativistic systems can be treated also. Discussions of the relativistic kinetic energy and exchange-correlation functionals have been included in chapters 8 and 9.

One of the earliest attemps to generalize the TF theory to the relativistic case was that of Vallarta and Rosen (37) with serious basic problems which remained unsolved even if efforts were made to correct them (38).

The density functional formalism itself was first generalized to include relativistic corrections by Rajagopal and Callaway (39) and given systematic derivation by MacDonald and Vosko (40) including the local density approximation. One of the new features of the relativistic treatments is the inclusion of a transverse exchange energy functional which can be more important than the relativistic corrections to the direct exchange part. A brief account of these matters in the case of atoms is given in chapter 8.

Some applications of density functional theory are mentioned in the different sections with the main purpose of illustrating the methods, techniques and the type of physical problems where they have been used.

REFERENCES

1. L.H. Thomas, Proc. Cambridge Philos. Soc. 23, 542 (1927).
2. E. Fermi, Z. Phys. 48, 73 (1928).
3. P.A.M. Dirac, Proc. Cambridge Philos Soc. 26, 376 (1930).
4. E. Wigner, Phys. Rev. 46, 1002 (1934).
5. C.F. v. Weizsacker, Z. Phys. 96, 431 (1935).
6. D.A. Kirzhnits, Zh. Ekop. Teor. Fiz. 32, 117 (1957); (Sov. Phys. - JETP 5, 64 (1957)).
7. H. B. Huntington, Phys. Rev. B 20, 3165 (1979).
8. P. Hohenberg and W. Kohn, Phys. Rev. B 136, 864 (1964).
9. M. Levi, Proc. Natl. Acad. Sci. (USA) 76, 6062 (1979).
10. L.J. Bartolotti, Phys. Rev. A 24, 1661 (1981).
11. S.C. Ying, Nuovo Cimento B 23, 270 (1974).
12. V. Peuckert, J. Phys. C 11, 4945 (1978).
13. A.K. Rajagopal, Adv. Chem. Phys. 41, 59 (1980).

14. W. Kohn and L.J. Sham, Phys. Rev. 140, A1133 (1965); L. Hedin and B.I. Lundqvist, J. Phys. C 4, 2064 (1971).

15. A.O. Amorin and R. Ferreira, Theoret. Chim. Acta (Berl.) 59, 551 (1981); R.G. Parr, R.A. Donelly, M. Levi, W.E. Palk Jr., J. Chem. Phys. 68, 3801 (1978); R.G. Parr, S.R. Gadre and L.J. Bartolotti, Proc. Natl. Acad. Sci. 76, 2522 (1979);L.J. Bartolotti, S.R. Gadre and R.G. Parr, J. Am. Chem. Soc. 102, 2945 (1980); R.P. Iczkowski and J.L. Margrave, J. Am. Chem. Soc. 83, 3547 (1961); E.P. Gyftopoulos and G.N. Hatsopoulos, Proc. Natl. Acad. Sci. 60, 786 (1968); N.H. March, Self-consistent fields in atoms, pp.44-45. Oxford: Pergamon Press 1975.

16. J.C. Slater, Quantum Theory of Molecules and Solids, Vol. 4, McGraw-Hill Book Company, New York 1974; A.S. Bamzai and B.M. Deb, Rev. Mod. Phys. 53, 95 (1981).

17. O. Gunnarson and B.I. Lundqvist, Phys. Rev. B 13, 4274 (1976).

18. V.L. Moruzzi, A.R. Williams and J.F. Janak, Phys. Rev. B 15, 2854 (1977).

19. U. von Barth and L. Hedin, J. Phys. C 5, 1629 (1972); O. Gunnarson, B.I. Lundqvist and J.W. Wilkins, Phys. Rev. B 10, 1319 (1974).

20. L. Eyges, Phys. Rev. 111, 683 (1958); K.H. Johnson, J. Chem. Phys. 45, 3085 (1966); K.H. Johnson, in Advances in Quantum Chemistry, edited by P.O. Löwdin, Vol. 7, pp. 143, Academic, New York, 1973; K.H. Johnson and F.C. Smith, Phys. Rev. B 5, 831 (1972); J. Keller, Int. J. Quantum Chem. 9, 583 (1975). Paper presented at the Sanibel Symposia (1973).

21. J. C. Slater, Phys. Rev. 81, 385 (1951).

22. J. Keller , Computational Methods for Large Molecules and Localized States in Solids, edited by F. Herman, A.D. McLean and R.K. Nesbet, 341-56, Plenum Press 1973 and J. Physique, 33, C3, 241 (1972); J. Keller, Hyperfine Interact. 6, 15 (1979).

23. J. Keller, J. Fritz and A. Garritz, J. Physique, 35, C4, 379 (1974);J. Keller and J. Fritz, Proceedings of the V Int. Conf. on Amorphous and Liquid Semiconductors 1973; A. Garritz and J. Keller, in Proceedings of the Int. Conf. on the Electronic and Magnetic Properties of Liquid Metals, University of México publications 1978.

24. M. Castro, J. Keller and P. Rius, Hyperfine Interactions 9, (1982). To be published.

25. J. Keller and M. Castro, J. of Magnetism and Magnetic Materials, 15-18, 856 (1980); T. Tanabe, H. Adachi and S. Imoto, Japan. J. Appl. Phys. 15, 1805 (1976); T. Tanabe, H Adachi and S. Imoto, Japan. J. Appl. Phys., 16, 1097 (1977); T. Tanabe, H. Adachi and S. Imoto, Japan. J. Appl. Phys. 17, 49 (1978); R.P. Messmer, D. R. Salahub, K.H. Johnson and C.Y. Yang, Chem. Phys. Letters 51, 84 (1977).

26. J. Avery and J.P. Dahl Editors, Local Densities in Quantum Chemistry and Solid State Theory, Plenum Press 1983.

27. J. Keller, C. Keller and C. Amador, Lectures Notes in Physics,

edited by J.G. Zabolitzky, M. de Llano, M. Fortes and J.W. Clark, 142, 364 (1981).

28. J.A. Alonso and L.A. Girifalco, Phys. Rev. B 17, 3735 (1978).

29. E. Fermi and E. Amaldi, Mem. Accad. Ital. 6, 117 (1934).

30. T.J. Tseng and M.A. Whitehead, Phys. Rev A 24, 21 (1981); T.J. Tseng and M.A. Whitehead, Phys. Rev. A 24, 16 (1981).

31. W. Kutzelnigg, G. del Re and G. Berthier, Phys. Rev. 172, 49 (1968).

32. M. S. Gopinathan, M.A. Whitehead and R. Bogdanović, Phys. Rev. A 14, 1 (1976).

33. K. Schwarz, Phys. Rev. 184, 10 (1969); for a review of the Xα method with additional references see, for example, J. C. Slater and J.H. Wood, Int. J. Quantum Chem. Symp. 4, 3 (1971).

34. J. Keller and J.L. Gázquez, Phys. Rev. A 20, 1289 (1979);J.L. Gázquez and J. Keller, Phys. Rev. A 16, 1385 (1977); J.L. Gázquez E. Ortiz and J. Keller, Int. J. of Quantum Chemistry, Quantum Chemistry Symposia, edited by P.O. Löwdin and Y. Ohrn 13, 377 (1979).

35. P.K. Acharya. L.J. Bartolotti, S.B. Sears and R.G. Parr, Proc. Natl. Acad. Sci. USA 77, 6978 (1980); J.L. Gázquez and J. Robles, J. Chem. Phys. 76, 1467 (1982).

36. M.S. Vallarta and N. Rosen, Phys. Rev. 41, 708 (1932).

37. E.K.U. Gross, A. Toepfer, B. Jacob and R.M. Dreizler, Proc. XVII Intern. Winter Meeting on Nuclear Physics (Bormio, 1979), Istituto Nazionale di Fisica Nucleare, pp. 84 (1979); M. Rudkjøbing, K. Dan Vidensk. Selsk. Mat. Fys. Medd. 27, No. 5 (1952); J.J. Gilvarry, Phys. Rev. 95, 71 (1954); N. Ashby and M.A. Holzman, Phys. Rev. A 1, 764 (1970).

38. A.K. Rajagopal and J. Callaway, Phys. Rev B 7, 1912 (1973); N. Eyashar and D.D. Koelling, Phys. Rev. B 15, 3620 (1977); D. Ellis, J. Phys. B: Atom. Molec. Phys. 10, 1 (1977).

39. A.H. MacDonald and S.H. Vosko, J. Phys C: Solid State Phys. 12, 2977 (1979); A.H. MacDonald, in Local Densities in Quantum Chemistry and Solid State Theory edited by J. Avery and J.P. Dahl, Plenum Press 1983.

THE CONSTRAINED SEARCH APPROACH, MAPPINGS TO EXTERNAL POTENTIALS, AND VIRIAL-LIKE THEOREMS FOR ELECTRON-DENSITY AND ONE-MATRIX ENERGY-FUNCTIONAL THEORIES

Mel Levy
Department of Chemistry and
Quantum Theory Group
Tulane University
New Orleans, Louisiana 70118

CONTENTS:

(Chapter for a textbook based upon the International Workshop on Electronic Density Functionals, University of Mexico, October 1980)

I. INTRODUCTION

Quantitative predictions by means of electronic wavefunctions, within the framework of the Schroedinger equation, continues to be quite cumbersome for systems large enough to be of interest because the dimensions of wavefunctions grow spacially as three times the number or electrons. Density functional theory provides an attractive alternative to wavefunctional theory because the electron density possesses only three dimensions no matter how large the system. Similarly, the reduced spacial one-matrix possesses only six dimensions regardless the size of the system. Furthermore, a formal justification for density functional theory arises from the fact, which is by now well-known, that a ground -state electron density contains implicitly all the information embedded within its ground-state wavefunction. Specifically, as proved by Hohenberg and Kohn, a ground-state electron density contains sufficient information to determine the more complicated ground-state wavefunction.[1]

In this chapter, we shall analyze the relationship between the ground-state density and the ground-state wavefunction, and the relationship between the ground-state density and its corresponding spin-free local external potential. We shall also discuss the universal variational functionals in density and one-matrix theories for ground-state energy calculations, and we shall present rigorous virial-like equalities and bounds. Finally, we shall compare wavefunction, two-matrix, one-matrix,[2-5] and Kohn-Sham formulations, and we shall discuss the appearance of noninteger occupation numbers within the Kohn-Sham theory.[6] In this connection, the classification of Slater's X-α theory,[7] the predecessor of formal Kohn-Sham theory, will be scrutinized as a one-matrix formulation. Featured throughout the chapter is the "constrained search" approach to density and one-matrix energy-functional theories.[5]

II. THE RELATIONSHIP BETWEEN A GROUND-STATE ELECTRON DENSITY AND ITS CORRESPONDING GROUND-STATE WAVEFUNCTION AND ENERGY

First of all, it is important to point out immediately that very many wavefunctions generally yield a given electron density. For instance, a single Slater determinant of spin-orbitals can always be found to yield the same density as the given ground-state wavefunction which is usually composed of an infinite number of Slater determinants. If only one wavefunction were to always yield a given ground-state density, then there would be no need for a Hohenberg-Kohn theorem because the existence of a ground-state density to ground-state wavefunction mapping would be trivially established. Namely, just find that wavefunction which yields the density. That wavefunction which fits this prescription would then automatically be identified as the ground-state wavefunction associated with the given ground-state density.

The realization that there is, in general, a many-to-one relationship between wavefunctions and a given ground-state density makes the problem more difficult, but the variational principle solves it for us by the following "constrained search" approach[5] to density functional theory:

Notice that all those antisymmetric wavefunctions that yield a given density possess the same expectation value with respect to any local external potential (the electron-nuclear attraction operator is an example of a local external potential). Consequently, of all those wavefunctions which yield the ground-state density, the ground-state wavefunction distinguishes itself as the one which minimizes the sum of the expectation values of the kinetic ($\hat{T}$) plus electron-electron repulsion ($\hat{V}_{ee}$) operators.

Hence, for the formal determination of the ground-state wavefunction from a ground-state electron density ρ, just compute the integral $< \Psi_\rho \mid \hat{T} + \hat{V}ee \mid \Psi_\rho >$ for each and every antisymmetric wavefunction Ψ_ρ which possesses the given ground-state density ρ. That Ψ_ρ which yields the minimum, Ψ_ρ^{min}, is the ground-state wavefunction associated with the given ρ. If degeneracies exist such that more than one Ψ_ρ gives this minimum, then all of the ground-state wavefunctions may be obtained, one at a time, by the above procedure. Furthermore, all of the wavefunctions which give the minimum must obviously be ground-statesof the same local external potential.

Once each Ψ_ρ^{min} has been determined, then by means of $\hat{H}\, \Psi_\rho^{min} = E\, \Psi_\rho^{min}$, it follows directly that the total multiplicative potential operator, $\hat{V}$, is obtained, within an additive constant, from Ψ_ρ^{min} by performing

$$\hat{V}-E = - \hat{T}\, \Psi_\rho^{min} / \Psi_\rho^{min} \tag{1}$$

Moreover, since $\hat{V}$ usually vanishes at infinity, E is given by

$$E = \lim_{\text{all } r_i \to \infty} \hat{T}\, (\vec{r}_1 \ldots \vec{r}_N)\; \Psi_\rho^{min}(\vec{r}_1 \ldots \vec{r}_N) \Big/ \Psi_\rho^{min}(\vec{r}_1 \ldots \vec{r}_N) \tag{2}$$

so that

$$\hat{V} = - \hat{T}\, \Psi_\rho^{min} / \Psi_\rho^{min} + \lim_{\text{all } \vec{r}_i \to \infty} \hat{T}\, (\vec{r}_1 \ldots \vec{r}_N)\; \Psi_\rho^{min}(\vec{r}_1 \ldots \vec{r}_N) \Big/ \Psi_\rho^{min}(\vec{r}_1 \ldots \vec{r}_N). \tag{3}$$

Finally, once $\hat{V}$ has been determined, the local external potential $\hat{v}$ may be obtained by

$$\sum_{j=1}^{N} \hat{v}(\vec{r}_j) = \hat{V} - \hat{V}ee \tag{4}$$

(Note that the spin in the N-electron wavefunction has been surpressed to simplify notation.)

So, by formal construction,[5] we have just witnessed how the ground-state density contains sufficient information to determine the gound-state wavefunction, the external potential, and the ground-state energy. In this section, we have emphasized the "constrained search" approach[5] which is a reformulation of the Hohenberg-Kohn orientation. We shall return to the "constrained search" approach when we discuss the variational methods.

III. THE MAPPING OF GROUND-STATE DENSITIES TO HAMILTONIANS [8]

Consider N interacting electrons in a local spin-independent external potential $\hat{v}$. (The external potential for a molecule consists of the electron-nuclear attraction operator, which is a coulomb potential, but it is important to note that we shall not necessarily restrict $\hat{v}$ to be a coulomb potential). The corresponding Hamiltonian is

$$\hat{H} = \hat{T} + \hat{V}ee + \sum_{j=1}^{N} \hat{v}(\vec{r}_j) \qquad (5)$$

The following assertion of the Hohenberg-Kohn thereom vividly illustrates the central role of the electron density in quantum chemistry and solid state physics: The ground-state density must change when the local external potential, $\hat{v}(\vec{r})$, changes by more than an additive constant.

We see that the marriage between density and external potential is an intimate one. In fact, in the last section it was shown quite explicitly how the external potential may be obtained, in a formal way, from its ground-state density. But, can one map a ground-state density to its external potential explicitly and exactly in a more practical manner? Well, yes and no. The best one can do (which is indeed aesthetically pleasing) is to map a set of M ground-state densities to a corresponding set of M Hamiltonians, where each density is associated with a different Hamiltonian in the form of Eq. (5). At the start, it is assumed that we do not know which density belongs with which $\hat{H}$. (Incidentally, it is important to note that for each coulomb $\hat{v}$ there are an infinite number of wrong densities with the right cusp conditions. In particular, a density might obey the electron-nuclear cusp conditions for a given coulomb $\hat{v}$ and yet be the ground-state density for some noncoulomb $\hat{v}$.) Following is the introduction to the theorem which achieves the mappings:

Label the M Hamiltonians $\hat{H}_1, \hat{H}_2, \ldots \hat{H}_M$ and the corresponding ground-state densities $\rho_1, \rho_2, \ldots \rho_M$, where ρ_i is the ground-state density of $\hat{H}_i$. The functional relationship between the H's and the ρ's shall be given by the study of

$$G^{\alpha,\beta,\ldots\omega}_{1,2,\ldots M} = \int d\vec{r}\, [\hat{v}_1(\vec{r})\rho_\alpha(\vec{r})+\hat{v}_2(\vec{r})\rho_\beta(\vec{r})+\ldots+ \hat{v}_M(\vec{r})\rho_\omega(\vec{r})] \tag{6}$$

where the superscripts in G are formed by permuting its subscripts. There are thus M! possible G's.

THEOREM: G achieves its unique minimum when $\alpha=1$, $\beta=2, \ldots \omega=M$.

Namely

$$\text{Min } G^{\alpha,\beta,\ldots\omega}_{1,2,\ldots M} = G^{1,2,\ldots N}_{1,2,\ldots N} = \int d\vec{r}[\sum_{i=1}^{M} \hat{v}_i(\vec{r})\rho_i(\vec{r})]. \tag{7}$$

PROOF: We begin along the lines of Eqs. (1-4) and Eq. (8) in reference 5c or along the lines of Eqs. (2.3-2.4) in reference 9 . As with the original Hohenberg-Kohn theorem, the existence of an inequality in Eq. (8) of reference 5c or in Eq. (2.4) of reference 9 dictates that $\hat{v}$ is a unique functional of ρ. But, Eq. (8) and Eq. (2.4) imply something much more as well. The direction of the inequality in these equations shall allow us to achieve our objective: By the variational theorem

$$\begin{aligned} &\langle\Psi_\alpha|\hat{H}_1|\Psi_\alpha\rangle + \langle\Psi_\beta|\hat{H}_2|\Psi_\beta\rangle + \ldots + \langle\Psi_\omega|\hat{H}_M|\Psi_\omega\rangle \\ &\geq \langle\Psi_1|\hat{H}_1|\Psi_1\rangle + \langle\Psi_2|\hat{H}_2|\Psi_2\rangle + \ldots + \langle\Psi_M|\hat{H}_M|\Psi_M\rangle \end{aligned} \tag{8}$$

where the Ψ's on the left-hand-side of Eq. (8) are permutations of the Ψ's on the right-hand-side. The equality clearly holds iff $\alpha=1$, $\beta=2$, ..., $\omega=M$. In any case, the sum of the $\langle\hat{T} + \hat{V}ee\rangle$'s on the left-hand-side must equal the sum of the $\langle\hat{T} + \hat{V}ee\rangle$'s on the right hand side. Hence it follows that

$$\begin{aligned} &\int d\vec{r}\,[\hat{v}_1(\vec{r})\rho_\alpha(\vec{r}) + \hat{v}_2(\vec{r})\rho_\beta(\vec{r}) + \ldots + \hat{v}_M(\vec{r})\rho_\omega(\vec{r})] \\ &\geq \int d\vec{r}\,[\hat{v}_1(\vec{r})\rho_1(\vec{r}) + \hat{v}_2(\vec{r})\rho_2(\vec{r}) + \ldots\hat{v}_M(\vec{r})\rho_M(\vec{r})] \end{aligned} \tag{9}$$

which is the desired result. That is,

$$G_{1,2,\ldots M}^{\alpha,\beta\ldots\omega} \leq G_{1,2,\ldots M}^{1,2,\ldots M}\,. \tag{10}$$

The equality obviously applies iff $\alpha=1$, $\beta = 2$, ..., $\omega=M$. The theorem is thus proved. Therefore, the ground-state ρ's are matched to their $\hat{H}$'s through the minimization of the functional $G_{1,2,\ldots M}^{\alpha,\beta,\ldots\omega}$ by the optimum ordering of the superscripts. The density denoted by the superscript is thereby the ground-state density for the external potential denoted by the corresponding subscript directly below the superscript in the optimum G.

IV. FORMAL UNIVERSAL VARIATIONAL FUNCTIONALS FOR $\langle \hat{T} + \hat{V}ee \rangle$

A. THE DENSITY

A variational principle involving the electron density was established indirectly by the original Hohenberg-Kohn theorem.[1] The existence of a proper universal functional for < T + Vee > of a trial ρ may be established quite directly by following in the spirit of the development in Section II. Without further delay, then, let us simply define the universal functional Q [ρ] as follows:[5]

$$Q[\rho] = \text{Min} \langle \Psi_\rho \mid \hat{T} + \hat{V}ee \mid \Psi_\rho \rangle. \quad (11)$$

Q [ρ] evaluates $\langle \hat{T} + \hat{V}ee \rangle$ for each and every wavefunction Ψ_ρ that yields the fixed trial ρ. The minimum in the expectation value search is the number assigned to Q [ρ]. (Even though wavefunctions are employed in the formal definition of Q [ρ], Q [ρ] is definitely a functional of ρ; input a ρ and Q [ρ] will output a number.) The variational principle, with ρ as a trial density for $\hat{v}$ (r), is established immediately. Just call Ψ_ρ^{min} that wavefunction which satisfies the search in Eq. (11). Then[5]

$$\int \hat{v}(\vec{r})\rho(\vec{r})\, d\vec{r} + Q[\rho] = \langle \Psi_\rho^{min} | \hat{H} | \Psi_\rho^{min} \rangle \geq E_v \quad (12)$$

where E_v is the ground-state energy for an $\hat{H}$ of form Eq. (2) with external potential $\hat{v}$. Furthermore, it is evident that the equality in Eq. (12) is achieved when the trial ρ is equal to the true ground-state density. Moreover, Parr, Donnelly, Levy, and Palke[10] have identified electronegativity with the Lagrange multiplier associated with the normalization constraint in Eq. (12).

Observe that Q [ρ] does not require that the trial ρ be a ground-state of some local external potential. In other words, unlike the indirect F [ρ] of Hohenberg and Kohn, v-representability[11] is not required in the definition of Q [ρ]; F [ρ], on the other hand, is defined only for v-representable ρ.

Specifically, F [ρ] is the $< \hat{T} + \hat{V}ee >$ of that ground-state wavefunction which yields ρ. In any case, F[ρ] = Q [ρ] when ρ is v-representable so that Q [ρ] is really an explicit formal display of the Hohenberg-Kohn functional F [ρ] which is only implicitly defined. Moreover, the following explicit formal display of a universal exchange-correlation functional, $E_{xc}[\rho]$, is consistent with Q [ρ], compatible with Kohn-Sham theory, and does not require ρ to be either interacting or non-interacting v-representable:[5b]

$$E_{xc}[\rho] = Q\,[\rho] - 1/2 \iint \rho(\vec{r}_1)\rho(\vec{r}_2) r_{12}^{-1} d\vec{r}_1 d\vec{r}_2 - \text{Min} < \Phi_\rho \mid \hat{T} \mid \Phi_\rho > \qquad (13)$$

where the Φ_ρ's are those antisymmetric wavefunctions which integrate to the trial ρ. In Section V we shall argue that fractional occupation numbers should result in the optimum equations of Kohn-Sham theory when the Φ_ρ^{min}, associated with the last term in Eq. (13), does not turn out to be a single determination.

In closing Part A, the reader should be made aware of the fact that Bartolotti[12] has recently utilized Eq. (11) in his time-dependent density functional theory which is based upon a hydrodynamic analogy to the Schroedinger equation, and Henderson[13] has put forth a counterpart of Q [ρ] for his density functional theory in momentum space. Finally, we note that excited-state density functional formulations have been reported by Theophilou,[14] Katriel,[15] Valone and Capitani,[16] and Levy.[17] It can be shown that Theophilou's formulation can be looked upon as a constrained ensemble search along the lines of Valone's ensemble extension[18] of the definition of Q [ρ] in Eq. (11).

B. THE ONE-MATRIX

The one-matrix (or first-order reduced density matrix)[19] contains more information than the density so that it is not surprising (using hindsight) that a variational principle exists,[2-4] for local external potentials, which involves the one-matrix, γ, and several groups have spearheaded work on universal variational functionals of γ for $\langle\hat{V}ee\rangle$. Specifically, Gilbert[2] established the existence of a functional of γ for a nonlocal external potential by extending the original Hohenberg-Kohn theorem to the nonlocal case (note that a local external potential may be looked upon as a special case of a non-local external potential). Berrondo and Goscinski,[3] with an orientation towards the implementation of Green's functions, added a non-local external potential to the N-body Hamiltonian in Eq. (5), and then derived a variational principle involving the one-matrix for a local external potential by eliminating the non-local external source. Donnelly and Parr[4] perceived that the original Hohenberg-Kohn theorem immediately implies the existence of a universal variational functional of γ. In particular, Donnelly and Parr utilized the fact that the Hohenberg-Kohn theorem implies that no two γ's may be identical if they belong to local external potentials which differ by more than an additive constant.

Given a pure-state N-representable[20,21] trial γ, the proper universal functional of γ for $\langle\hat{V}ee\rangle$ is[5]

$$W[\gamma] = \mathrm{Min}\,\langle\Psi_\gamma \mid \hat{V}ee \mid \Psi_\gamma\rangle \tag{14}$$

The search in Eq. (14) is conducted over all those antisymmetric wavefunctions which yield the fixed trial γ. With W [γ] thus defined, the variational principle follows at once:

$$E_v \leq \int dx[-1/2\nabla^2 + \hat{v}]\;\gamma(x'|x) + W\,[\gamma(x'|x)] \tag{15}$$

where x signifies space-spin coordinate. As with Q [ρ], it goes without saying that W[γ] has to be approximated in actual calculations.

For γ, the requisite of pure-state N-representability[21] is more severe than the requisite of ensemble N-representability.[20] If a γ is pure-state representable then it is automatically ensemble representable, but the converse is not always true. With this in mind, Valone, by taking into consideration Gilbert's analysis, has extended the search in Eq. (14) to include all those ensemble density matices which give the trial γ. Valone[18] has also expressed a reformulation with the reducing basis of Harriman.[22] For γ to be ensemble representable, it is necessary and sufficient that all the occupation numbers of γ be on the closed interval [o, 1] when γ is assumed to be normalized to the number of electrons N.[20]

Significantly, a meaningful approxiamtion to W[γ] has already been indirectly made. Lieb has very recently proven that when an ensemble representable γ is inserted into the Fock energy functional (which can be expressed entirely in terms of γ), the energy is rigorously bounded below by the Hartree-Fock energy, even when γ contains fractional occupation numbers.[23] Moreover, given a γ constructed from a certain one-electron basis set, it is entirely possible that the resultant energy could actually be lower than that of the best single determinant wavefunction constructed of orbitals expanded in this particular one-electron basis.[23]

V. COMPARISON OF ONE-MATRIX, KOHN-SHAM, AND SLATER'S X-α FUNCTIONAL THEORIES

By one-matrix functional theory, we shall mean a formulation which yields, upon energy minimization, that one matrix which arises from the exact interacting ground-state wavefunction:[2-5] Specifically, the ground-state γ is related to the ground-state Ψ by

$$\gamma(x'|x) = N \int \Psi^*(x', x_2, \ldots x_N)\, \Psi(x, x_2, \ldots x_N)\, dx_2 \ldots dx_N \tag{16}$$

Minimization of the right-hand-side of Eq. (15) yields the ground-state γ.

The γ in Eq. (16) can be expanded in terms of its natural spin-orbitals:[24]

$$\gamma\,(x'|x) = \sum_i d_i\, \lambda_i^*\,(x')\, \lambda_i\,(x) \tag{17}$$

The natural spin-orbitals, λ_i, are the eigenfunctions of γ and d_i are the corresponding eigenvalues.

We shall now discuss the formally exact Kohn-Sham theory in terms of the "constrained search" approach[5] to density functional theory, as exemplified by Eq. (11). Assume that a given trial ρ for an interacting system is simultaneously a ground-state density of some auxiliary noninteracting system. Then, the ground-state kinetic energy, which we shall call $T_q[\rho]$, has to be given by[5, 25a, 26]

$$T_q\,[\rho] = \mathrm{Min}\, \langle \Phi_\rho \,|\, \hat{T} \,|\, \Phi_\rho \rangle. \tag{18}$$

Kohn-Sham only considers ρ's which are noninteracting ground states, but this is not always the case. We shall, consequently, define $T_q\,[\rho]$ to be the kinetic energy of ρ even when ρ is not a noninteracting ground state.[5, 25a] Next, in the spirit of Kohn-Sham, we choose to put forth the following variational expression for an interacting system, with an even number of electrons, by taking into consideration Eq. (12):

$$E_v \le \int dx\, [-1/2\, \nabla^2 + \hat{v}]\, \tilde{\gamma}\,(x'|x) + \omega\,[\tilde{\gamma}], \tag{19}$$

where

$$\tilde{\gamma}\,(x'|x) = \sum_k n_k\, \phi_k^*\,(x')\, \phi_k\,(x) \quad ; \quad 0 \le n_k \le 1 \tag{20}$$

where [5b, 25a, 17]

$$\omega\,[\tilde{\gamma}] = Q\,[\rho] - T_q\,[\rho] \tag{21}$$

and where

$$\rho(\vec{r}) = \int \tilde{\gamma}\,(x|x)\,ds, \tag{22}$$

with s denoting spin.

The "pseudo" one-matrix $\tilde{\gamma}$ is assumed to be expanded in terms of the orthonormal spin-orbitals ϕ_k. To meet pure-state N-representability requirements for an even number of electrons, it is sufficient that the n_k in Eq. (20) be degenerate in pairs.[21]

The label "pseudo" has been attached to $\tilde{\gamma}$ because the optimum $\tilde{\gamma}$ should almost never turn out to be the same as the corresponding optimum γ. In fact, the optimum $\tilde{\gamma}$ would have to turn out to be idempotent (N occupied ϕ_k, each with $n_k = 1$) when the optimum ρ is a nondegenerate noninteracting ground-state density (noninteracting v-representable) because a noninteracting ground-state wavefunction is a single determinant. As implied above, noninteracting v-representability was assumed in the orginal Kohn-Sham paper[6] even though the authors did not concern themselves with the possibility of fractional occupation numbers. In any case, Eq. (19) certainly allows for the possibility of noninteger occupation numbers in the Kohn-Sham scheme, even though only integers should result in most instances. Exceptions include atomic multiplet problems when holes exist below the Fermi level. (For a related discussion concerning fractional occupation numbers consult Appendices A and B in the study by Perdew and Zunger.)[25a]

Even though the Kohn-Sham formulation does not yield the optimum γ, the formulation is nevertheless exact, in principle, in that the true ground-state energy and density are obtained. The X-α formulation of Slater[7] is actually a predecessor to the exact Kohn-Sham formulation as summarized in its general form by Eq. (19). In a way, the X-α scheme can be looked upon as either an approximation to Eq. (19) or to Eq. (15). I like to classify X-α as an approximation to Eq. (19) because the X-α occupation numbers often turn out to be unity in ground-state calculations, but strong arguments could be made for either classification.

Eq. (19) can be expressed as

$$E_v \leq \int dx \, [-1/2 \, \nabla^2 + \hat{v}] \, \tilde{\gamma} \, (x'|x) + 1/2 \iint \rho \, (\vec{r}_1) \rho \, (\vec{r}_2) \, | \, \vec{r}_1 - \vec{r}_2 |^{-1} d\vec{r}_1 d\vec{r}_2 + E_{xc}[\rho] \tag{23}$$

where [58,25a,17]

$$E_{xc} \, [\rho] = \omega \, [\tilde{\gamma}] \; - 1/2 \iint \rho(\vec{r}_1) \rho \, (\vec{r}_2) \, |\vec{r}_1 - \vec{r}_2|^{-1} \, d\vec{r}_1 d\vec{r}_2 \, . \tag{24}$$

In X-α theory, $E_{xc}[\rho]$ is approximately by the simple local functional*

$$E_{xc}[\rho] = \alpha \, C \int \rho \, (\vec{r})^{4/3} \, d\vec{r}, \tag{25}$$

where C is a constant and α is an adjustable paramenter. Much more elaborate approximations to $E_{xc}[\rho]$ are now, of course, in common use as discussed in various review articles and in other chaptersin this book. But, whatever approximations are used for $E_{xc}[\rho]$, they generally must mimic the constrained searches featured within this chapter.

Denote Ψ_v, γ_v, $\tilde{\gamma}_v$, and ρ_v as the ground-state functions of $\hat{H}$ with external

*For a completely local density functional theory, see references 27 and 28.

potential $\hat{v}$. According to the "constrained search" approach, the exact $E_{xc}[\rho_v]$ is not purely potential energy. Indeed, $E_{xc}[\rho]$ must contain positive kinetic energy because the search dictates that[5,17]

$$T_q[\rho_v] < \langle \Psi_v \mid \hat{T} \mid \Psi_v \rangle. \tag{26}$$

The magnitude of this positive kinetic energy, which equals $\langle \Psi_v \mid \hat{T} \mid \Psi_v \rangle$ $T_q[\rho_v]$, is expected to be of the order of magnitude of the correlation energy, as rationalized in the following development which exploits the method of integration over a coupling constant, a technique devised by Langreth and Perdew[29] and by Gunnarsson and Lundqvist[30] to elucidate the exchange-correlation hole.

Let us form a Hamiltonian analogous to Eq. (5), but now multiply $\hat{V}ee$ by a coupling constant λ:

$$\hat{H}(\lambda) = \hat{T} + \lambda \hat{V}ee + \sum_{j=1}^{N} \hat{v}(\lambda, \vec{r}_j). \tag{27}$$

For each λ $(0 \leq \lambda \leq 1)$, $\hat{v}(\lambda, \vec{r})$ is adjusted so that the ground-state density corresponding to $\lambda=1$, ρ_v, is always obtained, even though the ground-state wavefunction, $\Psi(\lambda)$, depends on λ (note that according to the notation $\Psi(\lambda=1)$ and Ψ_v are identical). Utilization of the Hellmann-Feynman theorem,

$$[\partial E(\lambda)/\partial\lambda] = \langle \Psi(\lambda) | \partial\hat{H}(\lambda)/\partial\lambda | \Psi(\lambda) \rangle, \tag{28}$$

followed by integration over λ, produces the appropriate formal expression for $E_{xc}[\rho_v]$:

$$E_{xc}[\rho_v] = \int_0^1 \langle \Psi(\lambda) | \hat{V}ee | \Psi(\lambda) \rangle$$
$$-1/2 \iint \rho_v(\vec{r}_1)\, \rho_v(\vec{r}_2)\, | \vec{r}_1 - \vec{r}_2 |^{-1} d\vec{r}_1\, d\vec{r}_2 \tag{29}$$

Implementation of the trapezoid rule as an approximation for the integration over λ on the right-hand-side of Eq. (29) gives[17]

$$E_{xc}[\rho_v] = 1/2\,[\langle\Psi(\lambda=1)\mid \hat{V}ee \mid(\Psi(\lambda=1)\rangle - \langle\Psi(\lambda=o)\mid \hat{V}ee \mid \Psi(\lambda=o)\rangle - \tag{30}$$

$$\iint \rho_v(\vec{r}_1)\rho_v(\vec{r}_2)\,\left|\vec{r}_1-\vec{r}_2\right|^{-1} d\vec{r}_1 d\vec{r}_2],$$

which in turn yields, after a few algebraic manipulations, the desired relationship:[17]

$$\langle\Psi_v \mid \hat{T} \mid \Psi_v\rangle - T_q[\rho_v] =$$

$$\langle\Psi(\lambda=o)\mid \hat{H}(\lambda=1)\mid \Psi(\lambda=o)\rangle - \langle\Psi(\lambda=1)\mid \hat{H}(\lambda=1)\mid \Psi(\lambda=1)\rangle = \mid E_{corr}\mid, \tag{31}$$

where E_{corr} signifies correlation energy. (By the way, notice that in density functional theory the Kohn-Sham single determinant often replaces the Hartree-Fock determinant as the basis for the definition of correlation energy.)

Perdew and Zunger[25a] have shown that $Q[\rho_v]$ as defined by Eq. (11), $T_q[\rho_v]$ as defined by Eq. (18), and $E_{xc}[\rho_v]$ as defined by Eq. (24) and expressed in Eq. (29), directly imply that the exchange-correlation energy of a single fully-occupied orbital in $\tilde{\gamma}_v$ must cancel exactly its self-direct coulomb energy. Partly with this in mind, Perdew and Zunger have created methods for incorporating a self-interaction corrections (SIC) into any density functional for the energy.[25] In particular, many of the anomalies of the local spin-density approximation have been removed by their SIC methods.

VI. VIRIAL-LIKE ENERGY-DENSITY AND ENERGY-ONE MATRIX RELATIONS

In the previous sections we discussed the formal variational functionals for computing ground-state energies and for the generation of the corresponding optimum densities and one-matrices. It should be evident that the external potential $\hat{v}(\vec{r})$ plays a very crucial role.[31] Once the desired optimum function (either density or one-matrix) is obtained, are there known exact simple relationships involving the optimum function, E_v, and $\hat{v}(\vec{r})$ which might serve as checks on the calculations? Well, there is no known operational energy-density equality.[32] We shall, however, present a fairly tight rigorous energy-density bound in this section. But first it is important to reveal, because it is not very well-known, that the exact ground-state electronic energy, for an interacting $\hat{H}$ of form Eq (5), can be expressed in terms of γ_v for any local external $\hat{v}(\vec{r})$, not just for a coulomb $\hat{v}(\vec{r})$, by means of the following exact and simple virial-like formula:[8]

$$E_v = 1/2 \int \nabla^2(\vec{r}) \; \gamma_v \; (x'|x)dX + \int d\vec{r}\hat{v}(\vec{r})\rho_v(\vec{r}) - \int d\vec{r}\rho_v \; (\vec{r}) \; [\partial\hat{v}(\lambda^{-1}\vec{r}/\partial\lambda]_{\text{at } \lambda=1} \; d\vec{r} \qquad (32)$$

where λ is a uniform scale factor of the electronic coordinates. The derivation of Eq. (32) follows along the usual lines of approach. Specifically, consideration is made of the fact that the ground-state wavefunction must be optimum with respect to a uniform scaling of its coordinates and consideration is made of the fact that $\hat{T}$ is homogeneous of degree - 2 and $\hat{V}ee$ is homogenous of degree - 1.

Let us now leave γ_v and focus entirely upon ρ_v to obtain our energy-density bound.[8] With this in mind, assume that the interacting ρ_v is simultaneously the ground-state density of the following auxiliary Kohn-Sham noninteracting Hamiltonian with the one-body potential $\hat{v}'$:

$$\hat{H}' = \hat{T} + \sum_{j=1}^{N} \hat{v}'(j). \tag{33}$$

Let the antisymmetric ground-state wavefunction of $\hat{H}'$ be called Ψ'. Then the homogeneity of $\hat{T}$ implies

$$<\Psi'|\hat{T}|\Psi'> = - 1/2 \int d\vec{r}\, \rho(\vec{r})\, [\partial \hat{v}'(\lambda^{-1}\vec{r})/\partial \lambda]_{\text{at } \lambda = 1} . \tag{34}$$

But

$$<\Psi'|\hat{T}|\Psi'> \leq - 1/2 \int \nabla^2(\vec{r}) \gamma_v\, (x'|x)\, dx \tag{35}$$

because Ψ' has to distinguish itself, from all the other antisymmetric wavefunctions which yield ρ_v, by minimizing the expectation value of $\hat{T}$ (it should be clear that $<\Psi'|\hat{T}\ |\Psi'> = T_q\ [\rho_v]$). Hence, from Eqs. (34) and (35)

$$1/2 \int \nabla^2(\vec{r})\gamma_v(x'|x)\, dx \leq 1/2 \int d\vec{r}\rho(\vec{r})\ [\partial \hat{v}(\lambda^{-1}\vec{r}) | \partial \lambda]_{\text{at } \lambda =1} . \tag{36}$$

Finally, when Eq. (36) is substituted into Eq. (32), the anticipated relation follows:[8]

$$E_v \leq \int d\vec{r}\hat{V}(\vec{r})\rho_v\ (\vec{r}) + 1/2 \int d\vec{r}\rho_v\ (\vec{r})\ [\partial \hat{v}'(\lambda^{-1}\vec{r}) | \partial \dot{\lambda}]_{\text{at } \lambda =1}$$

$$- \int d\vec{r}\ \rho_v(\vec{r})\ [\partial \hat{v}(\lambda^{-1} r)\ | \partial \lambda]_{\text{at } \lambda =1} . \tag{37}$$

The bound in Eq. (37) is expected to be reasonably tight. Indeed, the arguments in Section V dictate that the right-hand-side exactly exceeds E_v by the magnitude of the correlation energy when $<\Psi(\lambda)|\ \hat{V}ee\ |\Psi(\lambda)>$ is linear in λ, from $\lambda = o$ to $\lambda = 1$.

We end this section by reporting that semirigorous bounds are obtained if an N-electron atom is visualized as having been formed by gradual increases in the nuclear charge with simultaneous additions of electrons. This visualization presented by Levy and Tal leads to recursion relations involving E and $< r^{-1} >$. Specifically, given the $< r^{-1} >$'s for the first N atoms, E's for the first N atoms are obtained. One such recurson relation is [329,33]

$$E_N \leq -\sum_{k=1}^{N} (d\vec{r}r^{-1}\rho(K,H\text{-}F) + \varepsilon_o^K) \leq E_N^{H\text{-}F} \tag{38}$$

where E_K denotes the ground-state energy of a neutral atom with K electrons, E_K^{H-F} is the corresponding Hartree-Fock energy, $\rho(K,H\text{-}F)$ the corresponding Hartree-Fock density, and ε_o^K is the zero-order perturbation coefficient for K electrons (ε_o^K is known exactly from the one-electron atomic problem). The validity of the semirigorous bounds in Eq. (38) is supported through Ar, in Table I. The left-side inequality arises, in part, from the assumption that electron affinities are mostly positive and the right-side inequality arises, in part, from the assumption that the N-electron correlation energy is somewhat larger in magnitude than the corresponding sum, from K=1 to K=N, of electron affinities. In any case, the numbers from Eq. (38) always lie between the Hartree-Fock energies and the exact non-relativistic energies, and they generally fall closer to the latter.[329,33]

In concluding this section, it is important to note that quantum topology and subspace formulations have been developed in a promising manner to express the total electronic "energy density" as a functional of the one-matrix.[34]

VII. SOME CLOSING REMARKS

The full N-electron wavefunction increases in complexity quite rapidly as N grows. Alternatives to wavefunctional theory, therefore, have been sought for some time. The two-matrix, whose dimensionality does not grow with N, appeared for a while to be a marvelous possible alternative because the variational functional of the two-matrix is simple and known explicitly in terms of the reduced Hamiltonian. Fulfillment of the sufficient requisites for N-representability of a trial two-matrix is so difficult,[19,35] however, that there is now some pessimism concerning the future of direct variational calculations with the two-matrix. This brings us to one-matrix[2-5] and density[1] functional theories. For the one-matrix, ensemble N-representability is more easily satisfied than pure-state N-representability, but both are really little problem[20,21], and the density creates no problem at all.[2] In fact, a very specific prescription has recently been presented for the construction of a single Slater determinant of orthonormal orbitals from any density that is positive-definite and integrable.[36]

The universal variational functionals in one-matrix[2-4] and density[1] functional theories may be defined explicitly by means of the "constrained searches: as given by Eqs. (11) and (14).[5] Those functionals, of course, have to be approximated in actual calculations. What calculations have been carried out thus far? Well, while the number of calculations with $Q[\rho]$ implicitly in mind have been growing by leaps and bounds, through approximations to $E_{xc}[\rho]$, almost nothing has been done to approximate $W[\gamma]$ for the purpose of introducing electron-electron correlation, and one should study $W[\gamma]$ as well as $E_{xc}[\rho]$.

One advantage of $E_{xc}[\rho]$ over $W[\gamma]$ results from the fact that γ_v has to be more complex than ρ_v. On the other hand, $E_{xc}[\rho]$ is complicated by the fact that it must contain a positive kinetic energy contribution (of the order of the total correlation energy) as well as a large negative potential energy contribution. $W[\gamma]$, on the other hand, has to contain only pure potential energy. But then again, one can say that the necessary existence of a positive kinetic contribution to $E_{xc}[\rho]$ just means that approximations to $E_{xc}[\rho]$ have to provide that much less negative energy. Specifically, the correlation part of $W[\gamma]$ must be made to provide about twice as much negative energy as the correlation part of $E_{xc}[\rho]$.

Often when the expression "density functional theory" is employed, it is the formal Kohn-Sham theory and its variations which are automatically assumed. But perhaps Khon-Sham theory (of which $E_{xc}[\rho]$ is a part) is really a synthesis of a "pure density functional theory" and a "pure one-matrix functional theory" because Kohn-Sham generates, upon optimization, $\tilde{\gamma}_v$ and not γ_v; $\tilde{\gamma}_v$ is more complicated than ρ_v and less complicated than γ_v. In this connection, I have found in my discussions that many of us who have "grown up" with γ_v and natural spin-orbitals[24] are a bit skeptical of exact Kohn-Sham theory. We ask, how can correlation energy be obtained with an idempotent $\tilde{\gamma}_v$? Well, as reviewed in Section V in the present chapter, $\tilde{\gamma}_v$ does yield, in principle, the true ground-state energy, E_v, even though $\tilde{\gamma}_v$ is not the same as the true ground-state one-matrix, γ_v. On the other side of the coin, I feel that many of us who have "grown up" with Kohn-Sham theory, and are active in the us of $\tilde{\gamma}$, may not yet have full appreciation of the possibilities of γ, natural spin-orbitals, and $W[\gamma]$.[10, 2-5]

Finally, it gives me great pleasure to thank the organizers and participants for a most exciting workshop and to thank my collaborators on the works discussed within: Professors Robert G. Parr, Yoram Tal, John P. Perdew, William E. Palke, Robert Donnelly, and Stephen C. Clement.

REFERENCES

1. P. Hohenberg and W. Kohn, Phys. Rev. 136, B864 (1964).

2. T. L. Gilbert, Phys. Rev. B12, 2111 (1975).

3. M. Berrondo and O. Goscinski, Int. J. Quantum Chem. S9, 67 (1975).

4. R. A. Donnelly and R. G. Parr, J. Chem. Phys. 69, 4431 (1978); R. A. Donnelly, J. Chem. Phys. 71, 2874 (1979).

5. (a) M. Levy, Proc. Natl. Acad. Sci. USA 76, 6062 (1979).
   (b) M. Levy, Bull. Am. Phys. Soc. 24, 626 (1979).
   (c) M. Levy, J. Chem. Phys. 70, 1573 (1979).

6. W. Kohn and L. J. Sham, Phys. Rev. 140, A 1133 (1965).

7. J. C. Slater, The Self-Consistent Field for Molecules and Solids (McGraw-Hill, New York, 1974); J. W. D. Connolly in Semi-Empirical Methods of Electronic Structure Calculation, Part A, Plenum, N.Y., 1977.

8. Levy, Tal, and Clement (to be published).

9. P. W. Payne, J. Chem. Phys. 71, 490 (1979).

10. (a) R. G. Parr, R. A. Donnelly, M. Levy, and W. E. Palke, J. Chem. Phys. 68, 3801 (1978).
    (b) R. G. Parr, "Density Functional Theory of Atoms and Molecules", in Horizons of Quantum Chemistry, eds. K. Fukui and B. Pullman (D. Reidel, 1980).

11. Term coined by E. G. Larson

12. L. J. Bartolotti, Phys. Rev. A (in press).

13. G. Henderson, Phys. Rev. A 23, 19 (1981).

14. A. K. Theophilou, J. Phys. C 12, 5419 (1979).

15. J. Katriel, J. Phys. C 13, L375 (1980).

16. S. M. Valone and J. F. Capitani, Phys. Rev. A 23, 2127 (1981).

17. M. Levy, Phys. Rev. B (to be published).

18. S. M. Valone, J. Chem. Phys. 73, 1344, 4653 (1980).

19. For a comprehensive monograph, see E. R. Davidson, Reduced Density Matrices in Quantum Chemistry (Academic, New York, 1976).

REFERENCES

20. A. J. Coleman, Rev. Mod. Phys. 35, 668 (1963).

21. D. W. Smith, Phys. Rev. 147, 896 (1966).

22. J. Harriman, Phys. Rev. A 17, 1249, 1257 (1978). See also the chapters by Harriman in this book.

23. E. H. Lieb, Phys. Rev. Lett. 46, 457 (1981).

24. (a) P. O. Löwdin, Phys. Rev. 97, 1474 (1955).
    (b) P. O. Lowdin and H. Shull, Phys. Rev. 101, 1730 (1956).

25. (a) J. P. Perdew and A. Zunger, Phys. Rev. B 23, 5048 (1981).
    (b) A. Zunger, J. P. Perdew and G. L. Oliver, Solid State Commun. 34, 933 (1980).
    (c) J. P. Perdew, Chem. Phys. Lett. 64, 127 (1979).
    (d) A. Zunger and M. L. Cohen, Phys. Rev. B 18, 5449 (1978).

26. J. K. Percus, Int. J. Quant. Chem. 13, 89 (1978).

27. R. G. Parr, S. R. Gadre, and L. J. Bartolotti, Proc. Natl. Acad. Sci. USA 76, 2522 (1979).

28. S. R. Gadre, L. J. Bartolotti, and N. C. Handy, J. Chem. Phys. 72, 1034 (1980).

29. D. C. Langreth and J. P. Perdew, Phys. Rev. B 15, 2884 (1977).

30. O. Gunnarsson and B. I. Lundquist, Phys. Rev. B 13, 4274 (1976).

31. In this connection, see the chapters dealing with energies and electrostatic potentials at the nuclei in Chemical Applications of Atomic and Molecular Electrostatic Potentials, eds. P. Politzer and D. G. Truhlar (Plenum, New York, 1980).

32. There are, however, approximate relations.
    In particular, see
    (a) S. Frágá, Theoret. Chim. Acta. 2, 406 (1964).
    (b) R. Gaspar, Int. J. Quantum Chem. 1, 139 (1967).
    (c) P. Politzer and R. G. Parr, J. Chem. Phys. 61, 4258 (1974).
    (d) P. Politzer, J. Chem. Phys. 64, 4239 (1976).
    (e) Reference 8.
    (f) Reference 10a.
    (g) M. Levy and Y. Tal, J. Chem. Phys. 72, 3416 (1980).
    (h) M. Levy and Y. Tal, J. Chem. Phys. 73, 5168 (1980).
    (i) Y. Tal and M. Levy, Phys. Rev. A 23, 408 (1981).
    (j) Reference 5C.

REFERENCES

33. M. Levy, S. Clement, and Y. Tal, in Chemical Applications of Atomic and Molecular Electrostatic Potentials, eds. P. Politzer and D. G. Truhlar (Plenum, New York, 1980).

34. (a) R. F. W. Bader, J. Chem. Phys. 73, 2871 (1980) and references within.
(b) R. F. W. Bader, Y. Tal, S. E. Anderson, T. T. Nguyen-Dang, Israel J. Chem. 19, 8 (1980).
(c) S. Srebrenik, Int. J. Quant. Chem. S9, 375 (1975).

35. W. B. McRae and E. R. Davidson, J. Math. Phys. 13, 1527 (1972).

36. J. Harriman, "Orthonormal Orbitals for the Representation of an Arbitrary Density", WIS-TCI-606, 1980. (Theoretical Chemistry Institute, University of Wisconsin, Madison, Wisconsin).

37. P. O. Löwdin, Phys. Rev. 97, 1474 (1955).

38. P. O. Löwdin and H. Shull, Phys. Rev. 101, 1730 (1956).

39. G. C. Lie and E. Clementi, J. Chem. Phys. 60, 1275 (1974).

40. C. F. Fischer, The Hartree-Fock Method for Atoms, John Wiley and Sons, New York (1972).

TABLE I. Correlated Energies from Hartree-Fock Densities. (Energies are expressed in hartrees.)

Atom	Exact	Eq. (38)[c] (Hartree-Fock Densities)	Hartree-Fock Energy[d]
H	- 0.500	- 0.500	- 0.500
He	- 2.904	- 2.875	- 2.862
Li	- 7.477[a]	- 7.465	- 7.433
Be	- 14.666[a]	- 14.624	- 14.573
B	- 24.652[a]	- 24.628	- 24.529
C	- 37.842[a]	- 37.818	- 37.689
N	- 54.585[a]	- 54.529	- 54.401
O	- 75.061[a]	- 75.038	- 74.809
F	- 99.722[a]	- 99.682	- 99.409
Ne	- 128.93[b]	- 128.79	- 128.55
Na	- 162.24[b]	- 162.17	- 161.86
Mg	- 200.04[b]	- 199.98	- 199.61
Al	- 242.34[b]	- 242.31	- 241.88
Si	-289.35[b]	- 289.33	-288.85
P	-341.24[b]	- 341.20	-340.72
S	-398.10[b]	- 398.05	-397.50
Cl	-460.15[b]	-460.04	-459.48
Ar	-527.55[b]	-527.32	-526.82

[a] Obtained from reference 39.

[b] See page 88 in H. S. Schaefer III, "The Electron Structure of Atoms and Molecules" (Addison-Wesley, Reading, 1972). The uncertain Lamb correction has been ignored.

[c] Integrals are obtained from reference 40.

[d] Obtained from reference 40.

DENSITY MATRICES, REDUCED DENSITY MATRICES, A GEOMETRIC INVESTIGATION OF THEIR PROPERTIES, AND APPLICATIONS TO DENSITY FUNCTIONAL THEORY

John E. Harriman
Theoretical Chemistry Institute
University of Wisconsin-Madison
Madison, Wisconsin 53706

CONTENTS:

A series of three chapters. The first is a brief review of the properties of density matrices and reduced density matrices, including their spin properties. The second reviews the geometric approach to the theory of density matrices and the third applies this approach to density functional theory. These chapters have been prepared for inclusion in an advanced-level text book based on the International Workshop on Electronic Density Functionals held at the University of Mexico in October 1980. Chapters two and three are based on talks given at that workshop.

I. DENSITY MATRICES

A. Introduction

The intent of this chapter is to review briefly some of the more important properties of density matrices and reduced density matrices. In the interest of brevity, much will be omitted, and the interested reader is referred to other sources for more information.

B. Statistical Density Matrices

The density matrix was introduced into quantum mechanics by von Neumann (1927) [See also von Neumann (1955) or any book on quantum statistical mechanics.] to provide a way of treating statistical ensembles or systems in an incompletely-determined state. For a quantum mechanical system in a state described by a normalized wavefunction Ψ_k the average value of measurements of a property corresponding to a Hermitian operator $\hat{A}$ will be the expectation value

$$\langle\hat{A}\rangle_k = \int \Psi_k^* \hat{A} \Psi_k \, d\tau . \tag{1}$$

If it is known only that there is a probability p_k of a system being in state k, the appropriate average value is

$$\overline{\langle\hat{A}\rangle} = \sum_k p_k \langle\hat{A}\rangle_k . \tag{2}$$

The incompletely-characterized state or ensemble can be described by a <u>density matrix</u>

$$D(x_1,\ldots,x_n;x_1',\ldots,x_n') = \sum_k p_k \Psi_k(x_1,\ldots,x_n) \Psi_k^*(x_1',\ldots,x_n') \tag{3}$$

and the average expectation value given by

$$\overline{\langle\hat{A}\rangle} = \int dx_1\ldots dx_n \int dx_1'\ldots dx_n' \prod_{i=1}^{n} \delta(x_i-x_i')\hat{A}\, D(x_1,\ldots,x_n;x_1',\ldots,x_n')$$

$$\equiv \int' \hat{A}\, D(x_1,\ldots,x_n;x_1',\ldots,x_n')\, dx_1\ldots dx_n\ . \tag{4}$$

The operator $\hat{A}$ acts only on the unprimed variables in the density matrix D. In the first expression integration over the primed variables with the product of delta functions changes the primed variables into the corresponding unprimed variables after $\hat{A}$ has acted and the integration over unprimed variables then completes the evaluation. The second expression, with a prime on the integral sign, is introduced as a shorthand notation for this process. As we will see, the result can also be expressed as the trace of the product of operators $\hat{A}$ and $\hat{D}$. We have in mind a system of n electrons, and x_i stands for the space and spin variables associated with electron i.

An operator $\hat{D}$ acting on n-electron functions can be defined by

$$\hat{D}\,\Phi(x_1,\ldots,x_n) = \int D(x_1,\ldots,x_n;x_1',\ldots,x_n')$$

$$\times\ \Phi(x_1',\ldots,x_n')\, dx_1'\ldots dx_n' \tag{5}$$

and it follows from the definition that $\hat{D}$ is Hermitian. If a complete orthonormal basis set $\{\Phi_k\}$ is introduced, $\hat{D}$ can be represented by a matrix $\underset{\sim}{D}$ with

$$D_{k\ell} = \int \Phi_k^* \hat{D} \Phi_\ell \, dx_1 \ldots dx_n \tag{6}$$

and

$$D(x_1,\ldots,x_n;x_1',\ldots,x_n') = \sum_{k,\ell} D_{k\ell} \Phi_k(x_1,\ldots,x_n) \Phi_\ell^*(x_1',\ldots,x_n') \; . \tag{7}$$

The trace of $\hat{D}$ is by definition

$$\mathrm{tr}\,\hat{D} = \sum_k D_{kk} = \mathrm{tr}\,\underset{\sim}{D} \tag{8}$$

which is independent of the basis used. We will use the symbol D to mean the function D, the operator $\hat{D}$ or matrix $\underset{\sim}{D}$ as appropriate.

From Eq. (6) and the orthonormality of the basis,

$$\mathrm{tr}\,D = \int D(x_1,\ldots,x_n;x_1,\ldots,x_n) \, dx_1 \ldots dx_n \; . \tag{9}$$

Since the probabilities p_k must be such that

$$\sum_k p_k = 1 \; ; \qquad p_k \geqslant 0 \tag{10}$$

it follows that

$$\mathrm{tr}\, D = 1 \tag{11}$$

and that

$$D \geqslant 0 \,. \tag{12}$$

This last equation is the statement that D is positive semidefinite, i.e., any eigenvalue of D is nonnegative and for any function Φ

$$\int \Phi^* \hat{D}\, \Phi \, dx_1 \ldots dx_n \geqslant 0 \,. \tag{13}$$

A matrix can also be introduced corresponding to $\hat{A}$, and $\langle \hat{A} \rangle = \mathrm{tr}\, \underset{\sim}{AD}$.

Since the electronic wavefunctions Ψ or Φ are permutationally antisymmetric, $D(x_1,\ldots,x_n;x_1',\ldots,x_n')$ is antisymmetric with respect to permutations among the primed or among the unprimed indices.

We have defined D by Eq. (3) and obtained its properties. Alternatively, we could define a density operator, $\hat{D}$, as any positive Hermitian operator of trace 1. Such an operator is completely continuous (of trace class) so it has only discrete eigenvalues (no continuous region in its spectrum). The eigenvalues must be nonnegative and sum to 1, so the spectral expansion of D will be of the form given in Eq. (3).

A special case of a density matrix is the pure state, when one $p_k = 1$ and the remaining p_j's are 0. A density matrix describes a pure state if and only if

$$D^2 = D . \tag{14}$$

The operator D is then a projector onto the pure state Ψ_k.

C. Reduced Density Matrices

The idea of a reduced density matrix was introduced by Husimi (1940) and in the Hartree-Fock case by Dirac (1931). Reduced density matrices did not gain prominence in quantum chemistry and solid state physics, however, until the papers of Löwdin (1955) and McWeeny (1956, 1960).

Suppose that we have a many-electron system but are interested in the expectation value of a one-electron operator

$$\hat{F} = \sum_i \hat{f}(i) . \tag{15}$$

It will be given by

$$<\hat{F}> = \int' \hat{F}\, D\, dx_1 \ldots dx_n \tag{16}$$

with the primed integral convention defined earlier. (We need not be concerned with the distinction between a pure state and an ensemble at this point, so we drop the overbar indicating the ensemble average.) The operator $\hat{F}$ must be symmetric with respect to permutation of electron labels, since all electrons are the same, and D will be unchanged by any permutation applied to primed and unprimed indices alike. In each term of the sum Eq. (15) implies in Eq. (16), we can thus permute indices and relabel variables of integration to make that term the same as the first term. Thus

$$<\hat{F}> \;=\; n \int' \hat{f}(1)\; D(x_1,\ldots,x_n;x_1',\ldots,x_n')\; dx_1\ldots dx_n \;. \qquad (17)$$

The integration associated with variables $x_2 - x_n$ can be carried out independent of the operator $\hat{f}$, so we are led to define the <u>one-electron reduced density matrix</u>

$$D^{(1)}(x_1;x_1') \;=\; D(x_1,x_2,\ldots,x_n;x_1',x_2,\ldots,x_n)\; dx_2\ldots dx_n \qquad (18)$$

so that

$$<\hat{F}> \;=\; n \int' \hat{f}(1)\; D^{(1)}(x_1;x_1')dx_1 \;. \qquad (19)$$

A reduced density matrix (RDM) can be defined for any number of electrons less than n. Let $\underline{N}$ stand for the set of variables $x_1,\ldots,x_n$ and divide it into two subsets $\underline{N} = (\underline{P},\underline{Q})$ with $\underline{P} = x_1,\ldots,x_p$ and $\underline{Q} = x_{p+1},\ldots,x_n$. The p-electron RDM (also known as the p-th order RDM) is then defined as

$$D^{(p)}(\underline{P};\underline{P}') \;=\; \int D^{(n)}(\underline{P},\underline{Q};\underline{P}',\underline{Q})\; d\underline{Q} \qquad (20)$$

where we now write $D^{(n)}$ for the original, n-electron density matrix D, and $d\underline{Q} = dx_{p+1}\ldots dx_n$. For a p-electron operator

$$\hat{G} = \sum_{i_1<\cdots<i_p=1}^{n} \hat{g}(i_1,\ldots,i_p) \tag{21}$$

it follows that

$$<\hat{G}> = \binom{n}{p} \int' \hat{g}(1,2,\ldots,p)\ D^{(p)}(\underline{P};\underline{P}')d\underline{P}\ . \tag{22}$$

The factor $\binom{n}{p}$ can be absorbed into the density matrix, defining

$$\Gamma^{(p)}(\underline{P};\underline{P}') = \binom{n}{p}\ D^{(p)}(\underline{P};\underline{P}')\ . \tag{23}$$

In particular, Γ will be used for $\Gamma^{(2)}$ and γ for $\Gamma^{(1)}$ with this normalization.

A reduced density matrix can be thought of as describing an ensemble of p-electron states for a particular n-electron state or ensemble; it cannot be a p-electron "pure state" with only a single nonzero eigenvalue. In the Hartree-Fock case, however, the one electron RDM has n equal, nonzero eigenvalues and the rest are zero. Then γ is idempotent and is a projector onto the spinorbtial space spanned by the occupied spinorbitals.

Reduced density matrices can be very helpful in interpreting the behavior of many-electron systems, since they discard unnecessary information and allow us to concentrate our attention on the most relevant features. In this connection the natural spinorbitals which are the eigenfunctions of Γ and the natural spin geminals, which are the eigenfunctions of γ, have proved particularly useful. Some of these properties are discussed by Davidson (1972, 1976).

D. N-Representability

Given an n-electron Hamiltonian

$$\hat{H} = \sum_{i=1}^{n} f(i) + \sum_{\substack{i<j\\=1}}^{n} g(i,j) \tag{24}$$

we can define a reduced Hamiltonian

$$\hat{K} = \frac{1}{n-1}[f(1) + f(2)] + g(1,2) \tag{25}$$

and write the energy of a system with this Hamiltonian in terms of the two-electron RDM

$$E = \int' \hat{K}\, \Gamma(x_1,x_2;x_1',x_2')\, dx_1\, dx_2$$

$$= \binom{n}{2}\int' \hat{K}\, D^{(2)}(x_1x_2;x_1'x_2')\, dx_1\, dx_2\,. \tag{26}$$

Since $D^{(2)}$ contains most of the interesting information about the system and is a function of fewer variables than is the wavefunction for $n > 4$, it would be attractive to be able to determined $D^{(2)}$ directly, without going through a wavefunction. If this is simply done by varying a normalized $D^{(2)}$ in Eq. (26) so as to minimize E, unphysical results will be obtained unless appropriate constraints are imposed. Alternatively, one can start from the Schrödinger equation and obtain an equation which $D^{(2)} - D^{(4)}$ (or just $D^{(4)}$) must satisfy. In this case appropriate boundary conditions must be imposed if useful results are to be obtained. [Nakatsuji (1976), Cohen & Frishberg (1976), Harriman (1979)]

The problem of formulating completely these constraints or boundary conditions is the n-representability problem of reduced density matrix theory [Coleman (1963)]. It remains unsolved. It can be stated in general terms as follows: what are the conditions which must be imposed on a mathematical entity $D^{(p)}(\underline{P};\underline{P}')$ to assure that it can be obtained from some n-electron density matrix $D^{(n)}$ according to Eq. (20)? If it is further required that there be a pure-state density matrix from which $D^{(p)}$ can be obtained, the pure-state n-representability problem is defined.

The first condition to be noted is that a reduced density matrix is a density matrix: it must be Hermitian, positive, and of trace 1 (or some other fixed value such as $\binom{n}{p}$, depending on the normalization convention used). In addition, for $p > 1$ it must be antisymmetric with respect to permutations of the primed or the unprimed indices. If we denote the $\mathcal{P}^p$ the set of p-electron density matrices and by $\mathcal{P}_n^p$ the set of n-representable p-electron RDMs, then

$$\mathcal{P}_n^p \subset \mathcal{P}^p . \tag{27}$$

By considering reduction to proceed in steps, we find that

$$\mathcal{P}_n^p \subset \mathcal{P}_{n-1}^p \subset \cdots \subset \mathcal{P}_p^p = \mathcal{P}^p . \tag{28}$$

If a density matrix is n representable, it is also (n-1) representable, etc. Only for $D^{(n)}$ is it possible to have a pure state for an n-electron system.

The set P^n is convex, i.e., if U, $V \in P^n$ then

$$\alpha U + (1-\alpha)V \in P^n , \qquad 0 \leq \alpha \leq 1 . \qquad (29)$$

It follows from the linearity of the reduction process that P_n^p is also convex. Other topological properties of P^n and P_n^p can be obtained [e.g. Kummer (1967)], and many conditions which an element $D \in P_n^p$ must satisfy are known [e.g. Davidson (1969)].

In the case $p = 1$ the ensemble n-representability problem was solved by Coleman (1963): A density matrix $D^{(1)}$ is n representable if and only if its eigenvalues are all $\leq 1/n$. (Eigenvalues of γ must be ≤ 1.) D. W. Smith (1963, 1966) has shown that a sufficient condition for pure-state n-representability when n is even is that the Coleman condition be satisfied and that the eigenvalues be evenly degenerate. The problem for odd n is more difficult [Ruskai & Harriman (1968)] and has not been solved.

Although the n-representability problem has been most often discussed in topological terms, a geometric formulation of the properties of density matrices and n representability is possible when a finite basis set is introduced. [Harriman (1978)] This will be discussed in another chapter. One might wonder whether assuming a finite basis imposes a severe limitation. It does not in any practical sense for a molecular calculation, where a basis set expansion will be involved anyway in most methods, and we can consider the possibility of letting the basis set approach completeness. It might seem attractive to include r_{12} directly as a variable in $D^{(2)}$, since this approach has proved very effective in dealing with two-electron systems. This practical impossibility of doing so with exactly n-representable RDMs for n odd was demonstrated by Ruskai (1969) who showed that for odd n the rank (number of nonzero eigenvalues) of $D^{(2)}$ is finite if and only if the rank of $D^{(1)}$ is finite. Inclusion of r_{12} in $D^{(2)}$ implies an infinite rank for $D^{(1)}$, and we certainly want a finite-rank $D^{(2)}$. This means that $D^{(1)}$ must be a finite rank and each RDM, or the wavefunction for a pure state, can be expressed in terms of the finite set of spinorbitals associated with nonzero eigenvalues of $D^{(1)}$.

E. Spin Components

If one is concerned with systems in nonsinglet states, spin enters the theory of RDMs in a nontrivial way. Any RDM can be expressed as a sum of terms with each term a product of a spatial factor and a spin factor. These spin components have been extensively investigated for $D^{(1)}$ and $D^{(2)}$. [See, e.g., Bingel & Kutzelnigg (1970) or Davidson (1976) and references cited therein.] An extension to general p was recently developed [Harriman (1978c)] and the general formulation will be presented here.

The one-electron spin operators can be expressed in terms of ket-bra dyads in spin function space. We use irreducible tensorial components with a normalization that will be convenient later and write

$$[1]_1 = -\hat{s}_+ = -|\alpha\rangle\langle\beta|$$

$$[1]_0 = \sqrt{2}\,\hat{s}_z = \frac{1}{\sqrt{2}}(|\alpha\rangle\langle\alpha| - |\beta\rangle\langle\beta|)$$

$$[1]_{-1} = \hat{s}_- = |\beta\rangle\langle\beta| \; . \qquad (30)$$

We also want to deal with spin-free operators so we include something proportional to the spin-function-space identity operator

$$[s]_0 = \frac{1}{\sqrt{2}}\hat{1} = \frac{1}{\sqrt{2}}(|\alpha\rangle\langle\alpha| + |\beta\rangle\langle\beta|) \; . \qquad (31)$$

These operators can be combined to give irreducible tensorial spin operators for more than one electron: $\{p,j,u\}_m$ denotes component m of a rank j spin operator for p electrons produced by coupling scheme u. The coupling can be indicated schematically as shown in Figure 1. This diagram corresponds to the familiar branching diagram for the construction of spin eigenfunctions, and suggests another notation for the irreducible tensorial spin operators. An operator for electron p+1 can be combined with a p-electron operator of rank j in four ways: the rank-1 operators (Eq. (30)) can be introduced to produce a p+1 electron operator of rank j-1, j or j+1. These couplings will be dnoted by $\bar{1}$, 0, and 1, respectively. Alternatively, the rank 0 operator (Eq. (31)) can be introduced, necessarily leaving the total rank j. This will be denoted by s. Any irreducible tensorial spin operator is thus completely specified by a symbol such as $[1\bar{1}s10\ldots]_m$. The number of electrons involved is determined by the number of symbols

$\bar{1}$, 0, 1, or s; the rank j is the number of 1's minus the number of $\bar{1}$'s; and the coupling scheme is determined by the order of the symbols. The subscript m, with $-j \leqslant m \leqslant j$ again identifies a particular component.

For each p, j, and m there is one standard irreducible tensorial spin operator, $[1^j s^{p-j}]_m$ (i.e. in the symbol there are j 1's followed by p-j s's). In the diagram of Figure 1 this corresponds to moving along the left edge until the desired j is reached and then "hopping" to the right with unit operators to get the required p. Any irreducible tensorial spin operator can be expressed as a linear combination of permutations acting on the standard operator of the same p, j, and m.

The spin operators just defined provide a complete set, so that a density matrix can be expressed in terms of them with coefficients involving purely spatial variables.

$$D^{(p)}(x_1,\ldots,x_p;x_1',\ldots,x_p')$$

$$= \sum_{j=0}^{p} \sum_{m=-j}^{j} \sum_{u} T(\{p,j,j\}_m | \underline{r}_1,\ldots,\underline{r}_m;\underline{r}_1',\ldots,\underline{r}_p')\ \{p,j,u\}_m \ . \qquad (32)$$

Since each $\{p,j,u\}_m$ can be expressed in terms of $[1^j s^{p-j}]_m$, it follows from the overall permutational symmetry of $D^{(p)}$ that each spatial component $T(\{p,j,u\}_m)$ can be expressed in terms of permutations of the standard component $T([1^j s^{p-j}]_m)$.

If a density matrix describes an eigenstate of the n-electron spin operators $\hat{S}^2$ and $\hat{S}_z$, or is an ensemble of states all having the same values of the spin quantum numbers S and M, then it follows from the

Wigner-Eckart theorem that spin components differing only in m are simply related. This, together with the permutational relationships discussed above, means that only one component for each p and j need be considered independently.

When a density matrix is reduced, the integration (or partial trace) involves both space and spin variables. It follows from the orthonormality of the one-electron spin functions and the definitions introduced above that reduction from p to p-1 will produce 0 for any spin operator with $[\ ]_m$ symbol ending in $\bar{1}$, 0, or 1. As a consequence, the standard component with operator $[1^j s^{p-j}]_m$ will survive reduction longer than any other $\{p,j,u\}_m$. The standard component will contribute to $[1^j]_m$ in $D^{(j)}$ while any other component will be annihiliated by reduction before $D^{(j)}$ is reached.

The spin components of $\gamma(D^{(1)})$ and $\Gamma(D^{(2)})$ are summarized in Table 1.

TABLE 1. Spin Components of One- and Two-Electron Reduced Density Matrices.

One-Electron RDM

(The one-electron RDM has only standard components.)

$$D^{(1)} = T([s]_0)[s]_0 + T([1]_0)[1]_0$$

charge density matrix

$$[s]_0 = \frac{1}{\sqrt{2}}(|\alpha><\alpha| + |\beta><\beta|)$$

$$T([s]_0|\underline{r}_1;\underline{r}_1') = \frac{1}{\sqrt{2}}\int D^{(1)}(\underline{r}_1,\sigma_1;\underline{r}_1',\sigma_1)d\sigma_1$$

$$\gamma^0(\underline{r}_1;\underline{r}_1') = \sqrt{2}\,n\,T([s]_0|\underline{r}_1;\underline{r}_1')$$

spin density matrix (vanishes unless $S \geqslant 1/2$)

$$[1]_0 = \frac{1}{\sqrt{2}}(|\alpha><\alpha| - |\beta><\beta|)$$

$$T[1]_0|\underline{r}_1;\underline{r}_1') = \frac{1}{\sqrt{2}}\int' 2\hat{s}_{1z}\, D^{(1)}(\underline{r}_1,\sigma_1;\underline{r}_1',\sigma_1')d\sigma_1$$

$$\gamma^2(\underline{r}_1;\underline{r}_1') = \sqrt{2}\,n\,T([1]_0|\underline{r}_1;\underline{r}_1')$$

Two-Electron RDM

$$D^{(2)} = T([s^2]_0)[s^2]_0 + T([1\bar{1}]_0)[1\bar{1}]_0 + T([1s]_0)[1s]_0$$

$$+ T([s1]_0)[s1]_0 + T([10]_0)[10]_0 + T([1^2]_0)[1^2]_0$$

Table I continued

Standard Components

two-electron charge density

$$[s^2]_0 = \tfrac{1}{2}(|\alpha\alpha\rangle\langle\alpha\alpha| + |\alpha\beta\rangle\langle\alpha\beta| + |\beta\alpha\rangle\langle\beta\alpha| + |\beta\beta\rangle\langle\beta\beta|)$$

$$T([s^2]_0|\underline{r}_1,\underline{r}_2;\underline{r}_1',\underline{r}_2') = \tfrac{1}{2}\int D^{(2)}(\underline{r}_1,\sigma_1,\underline{r}_2,\sigma_2;\ \underline{r}_1',\sigma_1,\underline{r}_2',\sigma_2)d\sigma_1\ d\sigma_2$$

$$\Gamma^0(\underline{r}_1,\underline{r}_2;\underline{r}_1',\underline{r}_2') = 2\binom{n}{2}\ T([s^2]_0|\underline{r}_1,\underline{r}_2;\underline{r}_1',\underline{r}_2')$$

conditional spin density (vanishes unless $S \geqslant \tfrac{1}{2}$)

$$[1s]_0 = \tfrac{1}{2}(|\alpha\alpha\rangle\langle\alpha\alpha| + |\alpha\beta\rangle\langle\alpha\beta| - |\beta\alpha\rangle\langle\beta\alpha| - |\beta\beta\rangle\langle\beta\beta|)$$

$$T([1s]_0|\underline{r}_1,\underline{r}_2;\underline{r}_1',\underline{r}_2') = \tfrac{1}{2}\int' 2\hat{s}_{1z}\ D^{(2)}(\underline{r}_1,\sigma_1,\underline{r}_2,\sigma_2;\underline{r}_1',\sigma_1',\underline{r}_2',\sigma_2')d\sigma_1 d\sigma_2$$

$$\Gamma^z(\underline{r}_1,\underline{r}_2;\underline{r}_1',\underline{r}_2') = 2\binom{n}{2}\ T([1s]_0|\underline{r}_1,\underline{r}_2;\underline{r}_1',\underline{r}_2')$$

spin-spin coupling anisotropy (vanishes unless $S \geqslant 1$)

$$[1^2]_0 = \frac{1}{\sqrt{6}}(|\alpha\alpha\rangle\langle\alpha\alpha| - |\alpha\beta\rangle\langle\alpha\beta| - |\alpha\beta\rangle\langle\beta\alpha| - |\beta\alpha\rangle\langle\alpha\beta| - |\beta\alpha\rangle\langle\beta\alpha| + |\beta\beta\rangle\langle\beta\beta|)$$

$$T([1^2]_0|\underline{r}_1,\underline{r}_2;\underline{r}_1',\underline{r}_2') = \frac{1}{\sqrt{6}}\int' (3\hat{s}_{1z}\hat{s}_{2z} - \hat{\underline{s}}_1\cdot\hat{\underline{s}}_2)\ D^{(2)}(\underline{r}_1,\sigma_1,\underline{r}_2,\sigma_2;\underline{r}_1',\sigma_1',\underline{r}_2',\sigma_2')d\sigma_1\ d\sigma_2$$

$$\Gamma^{ss}(\underline{r}_1,\underline{r}_2;\underline{r}_1',\underline{r}_2') = \sqrt{6}\ \binom{n}{2}\ T([1^2]_0|\underline{r}_1,\underline{r}_2;\underline{r}_1',\underline{r}_2')$$

Table I concluded

Non-Standard Components

$$[1\bar{1}]_0 = \frac{1}{\sqrt{3}}(1 - \hat{P}_{1,2} - \hat{P}_{1',2'})[s^2]_0$$

$$T([1\bar{1}]_0) = \frac{1}{\sqrt{3}}(1 + \hat{P}_{1,2} + \hat{P}_{1',2'})\ T([s^2])_0$$

$$[s1]_m = \hat{P}_{1,2}\ \hat{P}_{1',2'}\ [1s]_m$$

$$T([s1]_m) = \hat{P}_{1,2}\ \hat{P}_{1',2'}\ T([1s]_m)$$

$$[10]_m = \frac{1}{\sqrt{2}}(\hat{P}_{1,2} - \hat{P}_{1',2'})\ [1s]_m$$

$$T([10]_m) = \frac{1}{\sqrt{2}}(-\hat{P}_{1,2} + \hat{P}_{1',2'})\ \ ([1s]_m)$$

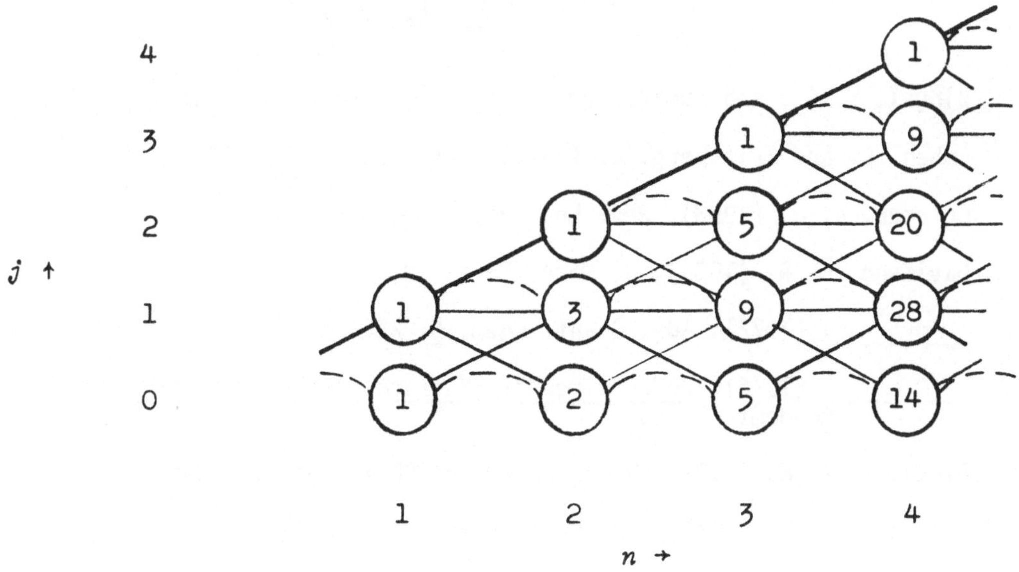

Figure 1. Coupling of Irreducible Tensorial Spin Operators. Operator ranks, j, for different numbers of electrons, n. The number in each circle indicates the number of distinct operators for each n, j, and m. It is equal to the number of ways that circle can be reached from the left and below. Solid lines indicate coupling with rank-1 one-electron operators and dashed lines indicate introduction of the unit operator of one-electron spin space.

References - Chapter I

Bingel, W. A., and Kutzelnigg, W. (1970), Adv. Quant. Chem. 5, 201.

Cohen, L. and Frishberg, C. (1976), Phys. Rev. A 13, 927.

Coleman, A. J. (1963), Rev. Mod. Phys. 35, 668.

Davidson, E. R. (1969), J. Math. Phys. 10, 725.

_________. (1972), Adv. Quant. Chem. 6, 235.

_________. (1976), Reduced Density Matrices in Quantum Chemistry, Academic Press.

Dirac, P. A. M. (1931), Proc. Cambridge Phil. Soc. 27, 240.

Harriman, J. E. (1978), (a) Phys. Rev. A 17, 1249; (b) ibid., 1257; (c) Internat. J. Quant. Chem. 15, 611.

_________. (1979), Phys. Rev. A 19, 1893.

Husimi, K. (1940), Proc. Phys. Math. Soc. Japan 22, 264.

Kummer, H. (1967), J. Math. Phys. 8, 2063.

Löwdin, P. O. (1955), Phys. Rev. 97, 1474.

McWeeny, R. (1956), (a) Proc. Roy. Soc. A235, 496; (b) ibid., A237, 335.

_________. (1960), Rev. Mod. Phys. 32, 335.

Nakatsuji, H. (1976), Phys. Rev. A 14, 41.

Ruskai, M. B. (1969), Phys. Rev. 183, 129.

Ruskai, M. B., and Harriman, J. E. (1968), Phys. Rev. 169, 101.

Smith, D. W. (1963), Rev. Mod. Phys. 35, 668.

_________. (1966), Phys. Rev. 147, 896.

von Neumann, J. (1927), Nachr. Akad. wiss. Göttingen, Math. physik. Kl IIa. Math. physik. chem. Abt. 1927, 245.

_________. (1955), Mathematical Foundations of Quantum Mechanics, transl. by R. T. Beyer, Princeton Univ. Press.

II. THE GEOMETRY OF DENSITY MATRICES

A. The Basic Geometric Formulation

Much of the formal theory of reduced density matrices has been developed in topological terms [e.g. Kummer (1967)]. A geometric formulation is possible, however, and will be summarized here [Harriman (1978)]. If the underlying quantum mechanics of a system of electrons is assumed to be approximated in terms of a finite basis set, density matrices become elements in a finite-dimensional vector space. Distances and geometric relationships in these spaces give additional insight into the properties of density matrices and the process of reduction, leading to reduced density matrices (RDMs).

It will be assumed that we have a finite, orthonormal spinorbital basis set $\{\phi_i, i = 1,\ldots,r\}$, with the number of functions r at least twice the maximum number of electrons we will want to deal with. We initially ignore any symmetry other than permutational. For any number, p, of electrons we define orthonormal basis functions as antisymmetrized products of the one-electron basis spinorbitals. With a ket notation,

$$|p,K\rangle = A\,\phi_{k_1}\cdots\phi_{k_p}\ , \qquad (1)$$

where $K = (k_1,\ldots,k_p)$ identifies the function. If a standard ordering is defined, e.g., $k_1 < k_2 < \ldots < k_p$, and K's ordered first by k_1, then by k_2, etc., K can also be considered as a single index with the range $1 \leq K \leq R_p = \binom{r}{p}$. Any quantum mechanical problem of interest (for a system of n electrons) is replaced by a model problem in which operators become finite matrices. For the Hamiltonian, for example,

$$\hat{H} = \sum_{i=1}^{n} \hat{f}(i) + \sum_{\substack{i<j \\ =1}}^{n} \hat{g}(i,j) \rightarrow \underset{\sim}{H}^{(n)}$$

$$H_{KL}^{(n)} = \langle n,K|\hat{H}|n,L\rangle \; . \qquad (2)$$

The "exact" solution of the Schrödinger equation for the model problem is thus the full CI result.

We begin by considering some fixed number of electrons and suppress the index p. The matrices of interest are of dimension $R = R_p = \binom{r}{p}$, and are Hermitian. Hermitian $R \times R$ matrices can be considered as elements in a real vector space. Clearly, a real linear combination of Hermitian matrices is a Hermitian matrix. With the trace scalar product

$$(\underset{\sim}{U},\underset{\sim}{V}) = \mathrm{tr}(\underset{\approx}{UV}) \qquad (3)$$

each element has a norm

$$\|\underset{\sim}{U}\| = \mathrm{tr}\,\underset{\sim}{U}^2 > 0 \qquad (4)$$

and the distance between $\underset{\sim}{U}$ and $\underset{\sim}{V}$ is

$$d_{\underset{\approx}{UV}} = \|\underset{\sim}{U} - \underset{\sim}{V}\| > 0 \; . \qquad (5)$$

This vector space will be denoted by E. Since each $R \times R$ Hermitian matrix is determined by R^2 real numbers (R real diagonal elements and

½R(R-1) independent, complex off-diagonal elements) we can construct R^2 linearly independent matrices: the dimension of E is R^2.

A simple basis set for E can be constructed as

$$\underset{\sim}{A}^{K} \leftrightarrow |K\rangle\langle K| \,, \qquad A^{K}_{IJ} = \delta_{IK}\delta_{JK}$$

$$\underset{\sim}{B}^{KL} \leftrightarrow \frac{1}{\sqrt{2}}(|K\rangle\langle L| + |L\rangle\langle K|)$$

$$B^{KL}_{IJ} = \frac{1}{\sqrt{2}}(\delta_{IK}\delta_{JL} + \delta_{IL}\delta_{JK})$$

$$\underset{\sim}{C}^{KL} \leftrightarrow \frac{i}{\sqrt{2}}(|K\rangle\langle L| - |L\rangle\langle K|)$$

$$C^{KL}_{IJ} = \frac{i}{\sqrt{2}}(\delta_{IK}\delta_{JL} - \delta_{IL}\delta_{JK}) \tag{6}$$

for $1 \leqslant K < L \leqslant R$. If we limit ourselves to real symmetric matrices a space of only $R(R+1)/2$ dimensions is involved and the c-type basis elements can be omitted. It is readily verified that all these basis elements are orthonormal with respect to the trace scalar product.

The trace of an individual matrix will also play a role in our development, and it is a linear property so that the trace of any element will be a linear combination of the traces of the basis elements. It follows immediately from the definitions that

$$\mathrm{tr}\, \underset{\sim}{B}^{KL} = \mathrm{tr}\, \underset{\sim}{C}^{KL} = 0$$

$$\mathrm{tr}\, \underset{\sim}{A}^{K} = 1 \,. \tag{7}$$

We now change to a new basis in which the B's and C's remain unchanged but the A's are replaced by $\bar{A}$'s with

$$\underset{\sim}{\bar{A}}^0 = R^{-1/2} \sum_{K=1}^{R} \underset{\sim}{A}^K \tag{8}$$

and $\underset{\sim}{\bar{A}}^1,\ldots,\underset{\sim}{\bar{A}}^{R-1}$ required to be normalized and orthogonal to $\underset{\sim}{\bar{A}}^0$ and each other, but otherwise not now specified. A particular choice of these basis elements, togther with a change of the $\underset{\sim}{B}$ and $\underset{\sim}{C}$ basis elements, will be considered later.

Suppose that the transformation from $\underset{\sim}{A}$'s to $\underset{\sim}{\bar{A}}$'s is

$$\underset{\sim}{\bar{A}}^K = \sum_{L=1}^{R} C_{KL}\, \underset{\sim}{A}^L , \qquad K = 0,1,\ldots,R-1 \tag{9}$$

with $C_{0,L} = R^{-1/2}$. Then

$$\text{tr}\, \underset{\sim}{\bar{A}}^K = \sum_L C_{KL} \tag{10}$$

and $\text{tr}\, \underset{\sim}{\bar{A}}^0 = R \cdot R^{-1/2} = R^{1/2}$. We have required the other $\underset{\sim}{\bar{A}}^K$ to be orthogonal to $\underset{\sim}{\bar{A}}^0$, so

$$\begin{aligned} 0 = (\underset{\sim}{\bar{A}}^0,\underset{\sim}{\bar{A}}^K) &= \sum_{J=1}^{R} \sum_{L=1}^{R} C_{0J}\, C_{KL}\, (\underset{\sim}{A}^J,\underset{\sim}{A}^L) \\ &= \sum_{L=1}^{R} R^{-1/2}\, C_{KL} = R^{-1/2}\, \text{tr}\, \underset{\sim}{\bar{A}}^K \end{aligned} \tag{11}$$

and thus $\text{tr}\, \underset{\sim}{\bar{A}}^K = 0$ for $K \neq 0$. In this new basis $\underset{\sim}{\bar{A}}^0$ is the only element with a nonzero trace and the trace of any element in the space will be just $R^{1/2}$ times the coefficient of $\bar{\underset{\sim}{A}}^0$ in the expansion of that element in terms of the basis.

Let *N* be the space spanned by the basis elements other than $\bar{\underset{\sim}{A}}^0$. It is the space of trace-zero Hermitian R R matrices. It is convenient to define

$$\underset{\sim}{X} = R^{-1/2} \bar{\underset{\sim}{A}}^0 = R^{-1} \sum_{K=1}^{R} \underset{\sim}{A}^K = R^{-1} \underset{\sim}{1} \qquad (12)$$

so that $\text{tr}\, \underset{\sim}{X} = 1$. Then any matrix V in *E* can be expressed as

$$\underset{\sim}{V} = (\text{tr}\, \underset{\sim}{V})\underset{\sim}{X} + \underset{\sim}{V}' \qquad (13)$$

where $\underset{\sim}{V}' = \underset{\sim}{V} - (\text{tr}\, \underset{\sim}{V})\underset{\sim}{X}$ is an element of *N* . Any density matrix (which must be of trace 1) is expressible as $\underset{\sim}{X}$ plus an element of *N*.

B. N-Matrices and 1-Matrices

With the preliminaries developed above we are ready to consider some of the geometric properties of the space of density matrices. They will be presented here without proof, since proofs are available elsewhere [Harriman (1978a)]. In such a development figures can be useful as an aid to understanding and as mnemonics. They may also suggest properties of, or relationships among, the quantities of interest, but any such result must be verified by algebraic or rigorous geometric proof. In a space of a large number of dimensions, such as we have here, even Euclidean geometry may differ significantly from our intuitive ideas developed in two and three dimensions.

In addition to having unit trace, a density matrix must be positive. This is a much more difficult condition to deal with, since it is exceedingly nonlinear. Indeed, a sufficiently convenient geometric characterization of the space of positive operators would in effect provide a solution to the n-representability problem. We don't have it yet.

We define a subset P of E as the set of $R \times R$ Hermitian matrices which are positive, and a subset P' of N as the set of matrices $\underset{\sim}{D}' \in N$ such that $\underset{\sim}{D} = \underset{\sim}{D}' + \underset{\sim}{X} \in P$ and thus $\underset{\sim}{D}$ is a density matrix. As a partial characterization of this space we define in N two concentric hyperspheres about the origin. (The origin in N corresponds to the point $\underset{\sim}{X}$ in E.) The outer hypersphere is of radius $[(R-1)/R]^{1/2}$ and the inner hypersphere is of radius $[R(R-1)]^{-1/2}$. The boundary of P' lies between these two hyerspheres and intersects each of them (in the sense of tangency of two hypersurfaces over a subspace of lower dimensionality). The situation is illustrated schematically in Figure 1.

The location of the boundary of P' between these hyperspheres means that for any $\underset{\sim}{D}'$ within the inner hypersphere, $\underset{\sim}{D}' + \underset{\sim}{X}$ is a density matrix, and for any density matrix $\underset{\sim}{D}$, $\underset{\sim}{D} - \underset{\sim}{X}$ lies within the inner hypersphere. The intersection of the boundary of P' with the outer hypersphere is precisely the space whose elements, when added to $\underset{\sim}{X}$, are pure state density matrices. Directly opposite, in the sense of travel along a straight line through the origin, any point where the boundary of P' intersects the outer hypersphere is a point where it intersects the inner hypersphere.

We turn next to the special case of one-electron density matrices (p=1). Since the ensemble n-representability problem for one-matrices has been solved [Coleman (1963)], we can investigate the properties of the space P_n^1 of n-representable one-electron RDMs, and the corresponding space $P_n'^1$ with $\underset{\sim}{X}$ subtracted from each RDM. This situation is represented schematically in Figure 2. In addition to the inner and outer hyperspheres defined previously, with now $R = R_1 = \binom{r}{1} = r$, we define for an intermediate hypersphere of radius $[(r-n)/rn]^{1/2}$ dependent on n which contains $P_n'^1$. The intersection of this hypersphere with the boundary of $P_n'^1$ consists of those $\underset{\sim}{D}^{(1)} - \underset{\sim}{X}^{(1)}$ for which $\underset{\sim}{D}^{(1)}$ is a Slater one-matrix: i.e. $\underset{\sim}{\gamma} = n\,\underset{\sim}{D}^{(1)}$ is idempotent and $\underset{\sim}{D}^{(1)}$ is n representable by a single determinant. The inner hypersphere plays the same role as before, and any point within it corresponds not only to a density matrix but also to an n-representable RDM for any n. (Clearly $n \leqslant r$, but we initially assumed $r \geqslant 2n$ for any n of interest.) It is not true, however, that opposite any intersection of $P_n'^1$ with the intermediate hypersphere is an intersection with the inner hypersphere.

C. Reduced Density Matrices and the Reducing Basis

We have seen some of the geometric characteristics of the space P^p of p-electron density matrices and of the space P_n^1 of n-representable one-matrices. We would like to use this knowledge to get information about the spaces of n-representable p-matrices with $1 < p < n$. In doing so we make use of the fact that the reduction operation is a linear mapping from one space to another. We will see that great simplifications arise from the choice of a particular basis set in each space. [Harriman (1978b)]

We now use E_p to denote the space of $\binom{r}{p} \times \binom{r}{p}$ Hermitian matrices defined with respect to the basis functions $\{|p,K\rangle\}$ defined in Eq. (1), and consider the space

$$E = E_1 \oplus E_2 \oplus \cdots \oplus E_n \oplus \cdots \oplus E_r . \qquad (14)$$

This space is related to the individual spaces E_p in the same way that Fock space is related to the Hilbert spaces for various numbers of electrons. A general element of E is expressible as a sum of components in the different E_p's. Within E we define a linear operator corresponding to reduction

$$\underset{\sim}{D}^{(q)} = \hat{L}_p^q \underset{\sim}{D}^{(p)} \qquad q < p \qquad (15)$$

and use a special symbol for the operation of reduction by 1

$$\hat{\Lambda}_- = \binom{p}{p-1} \hat{L}_p^{p-1} = p\,\hat{L}_p^{p-1} . \qquad (16)$$

Another operator, $\hat{\Lambda}_+$, can be defined by requiring that for each $\underset{\sim}{D}^{(p)} \in E_p$, for any p, $\hat{\Lambda}_+ D^{(p)}$ = the sum of all elements in E_{p+1} which reduce to $\underset{\sim}{D}^{(p)}$ on the application of $\hat{L}_{p+1}^p$. [A precise definition is given elsewhere, Harriman (1978b).] Three further operators are then defined as

$$\hat{\Lambda}_1 = \frac{1}{2}(\hat{\Lambda}_+ + \hat{\Lambda}_-)$$

$$\hat{\Lambda}_2 = -\frac{i}{2}(\hat{\Lambda}_+ - \hat{\Lambda}_-)$$

$$\hat{\Lambda}_3 = \frac{1}{2}(\hat{\Lambda}_+\hat{\Lambda}_- - \hat{\Lambda}_-\hat{\Lambda}_+) \ . \tag{17}$$

It can be shown that these operators obey the angular momentum commutation relationships

$$[\hat{\Lambda}_j, \hat{\Lambda}_k] = i\,\hat{\Lambda}_\ell \tag{18}$$

where j, k, ℓ is any cyclic permutation of 1,2,3, and that the eigenvalues of $\hat{\Lambda}_3$ determine the number of particles, p. (The eigenvalue of $\hat{\Lambda}_3$ for any element of E_p is (2p-r)/2 .) Of course any of these $\hat{\Lambda}_j$'s commutes with

$$\hat{\Lambda}^2 = \hat{\Lambda}_1{}^2 + \hat{\Lambda}_2{}^2 + \hat{\Lambda}_3{}^2 \tag{19}$$

corresponding to the square of the total angular momentum.

An analogy which is essentially an isomorphism can be established between the basis elements for E and spin product functions. If a transformation is made to eigenfunctions of Λ^2 (analogous to $\hat{S}^2$ eigenfunctions), then the eigenvalue and coupling scheme provide properties which are invariant under reduction. Reduction corresponds to applying $\hat{S}_-$, and at the point corresponding to $M_S = -S$ an attempt at further reduction will give zero. It is in this way that the dimensionality of the spaces E_p decreases naturally with decreasing p. The basis in each E_p of $\hat{\Lambda}^2$ eigenfunctions (with the same coupling scheme for each p) is referred to as the reducing basis.

The properties of the reducing basis can be summarized as follows: The reducing basis elements of E_p can be divided into subsets (each subset spanning a subspace of E_p) characterized by an index π. If the reduction operator $\hat{L}_p^q$ is applied to a reducing basis element of E_p, the result is either 0 (if $q < \pi$) or proportional to a reducing basis element of (if $\pi \geq q$). The basis element in E_q will be labeled by the same value of π and coupling-scheme index as the initial basis element in E_p. Given any element $\underset{\sim}{D}^{(q)}$ in E_q, its preimage in E_p (the set of all elements in E_p which reduce to $\underset{\sim}{D}^{(q)}$) can be correspondingly resolved into two orthogonal subsets: The component of the preimage in one of them is completely determined by $\underset{\sim}{D}^{(q)}$, the component in the other is completely free.

D. Spin Components

Each of the density matrices can be resolved into spin components in a way which is particularly useful when one is interested in RDMs corresponding to n-electron $\hat{S}^2$, $\hat{S}_z$ eigenstates, or ensembles of such states. A basis set of orbitals is introduced and the spinorbitals expressed as products of these orbitals and the two spin functions for one electron. When spin and spatial dependence is separated in the RDMs, the permutational symmetry is complicated. It is nevertheless possible to introduce basis functions with the appropriate permutational symmetry for each spin component and each number of electrons.

Spatial density matrix components are determined by matrices in the appropriate basis sets, and reduction properties can be treated by techniques like those considered above. [Harriman (1978c)] Since each irreducible tensorial spin operator is either annihilated by reduction or reduces to an irreducible tensorial spin operator for the smaller number of electrons, a series of spaces for a particular set of components are related.

We consider in particular the two-electron charge density matrix. The appropriate basis functions in this case are symmetrized or antisymmetrized products of orbitals, and the two subspaces are noninteracting, i.e. there are no nonzero matrix elements of the two electron charge density matrix between symmetric and antisymmetric basis functions. [Kutzelnigg (1963), (1965)] The density matrix component can thus be resolved into symmetric and antisymmetric pieces, and each piece can be separately reduced to a one electron density matrix spatial component. In nonbasis-set terms,

$$\Gamma^0 = \Gamma^a + \Gamma^s$$

$$\gamma^a(\underline{r}_1;\underline{r}_1') = \frac{2}{n-1}\int \Gamma^a(\underline{r}_1,\underline{r}_2;\underline{r}_1',\underline{r}_2)\, d\underline{r}_2$$

$$\gamma^s(\underline{r}_1;\underline{r}_1') = \frac{2}{n-1}\int \Gamma^s(\underline{r}_1,\underline{r}_2;\underline{r}_1',\underline{r}_2)\, d\underline{r}_2 \ . \qquad (20)$$

The total two-electron charge density matrix reduces to the charge density matrix

$$\gamma^0(\underline{r}_1;\underline{r}_1') = \frac{2}{n-1}\int \Gamma^0(\underline{r}_1,\underline{r}_2;\underline{r}_1',\underline{r}_2)\, d\underline{r}_2 = \gamma^a + \gamma^s . \quad (21)$$

The spin density matrix is the reduction of the conditional spin density matrix, which is an independent component of $\underset{\sim}{D}^{(2)}$. Nevertheless, it can be shown that

$$\gamma^z = \frac{n+2}{S+1}\gamma^a - \frac{3(n-2)}{S+1}\gamma^s . \quad (22)$$

This means that the spin density matrix is in fact determined by the two-electron charge density matrix (for $\hat{S}^2$, $\hat{S}_z$ eigenstates or ensembles of states with the same spin quantum numbers).

The geometric characterizations of DMs and RDMs described above provide geometric properties of the spin components are well, some of which have been investigated.

E. Conclusions

The geometric description of density matrices and RDMs provides a way of understanding some of their properties and places constraints on acceptable density matrices. The reducing basis is particularly useful in a consideration of RDMs, and allows us to determined, for example, precisely what it means to constrain a wavefunction or n-matrix to give a particular RDM on reduction. This is relevant to the theory of density functionals, particularly with the Levy algorithm. [Levy (1979)] It can be applied directly when density functional theory is expressed in terms of the one-electron RDM. The relationship to the density will be determined by the relationship between the density and the one-electron RDM, which will be discussed in the next chapter.

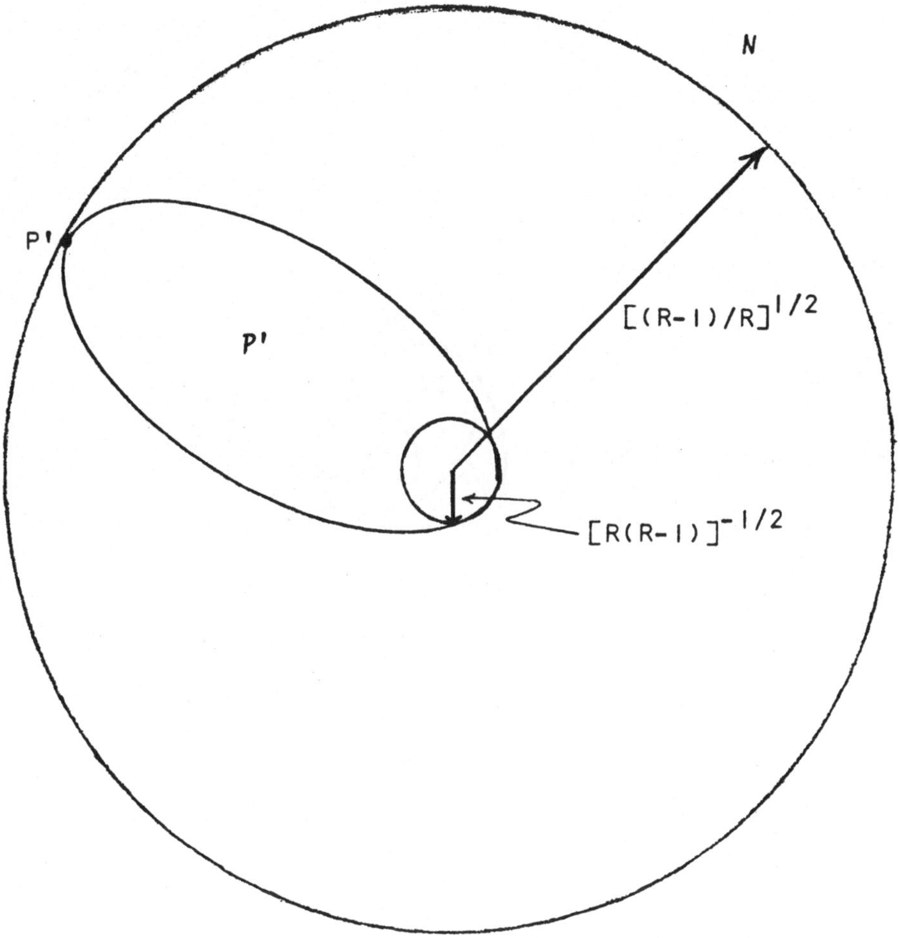

Figure 1. The Subset *P'* in *N*.

The subset corresponding to positive operators of trace 1 has a boundary lying between two hyperspheres. The point of intersection P' corresponds to a pure state density matrix.

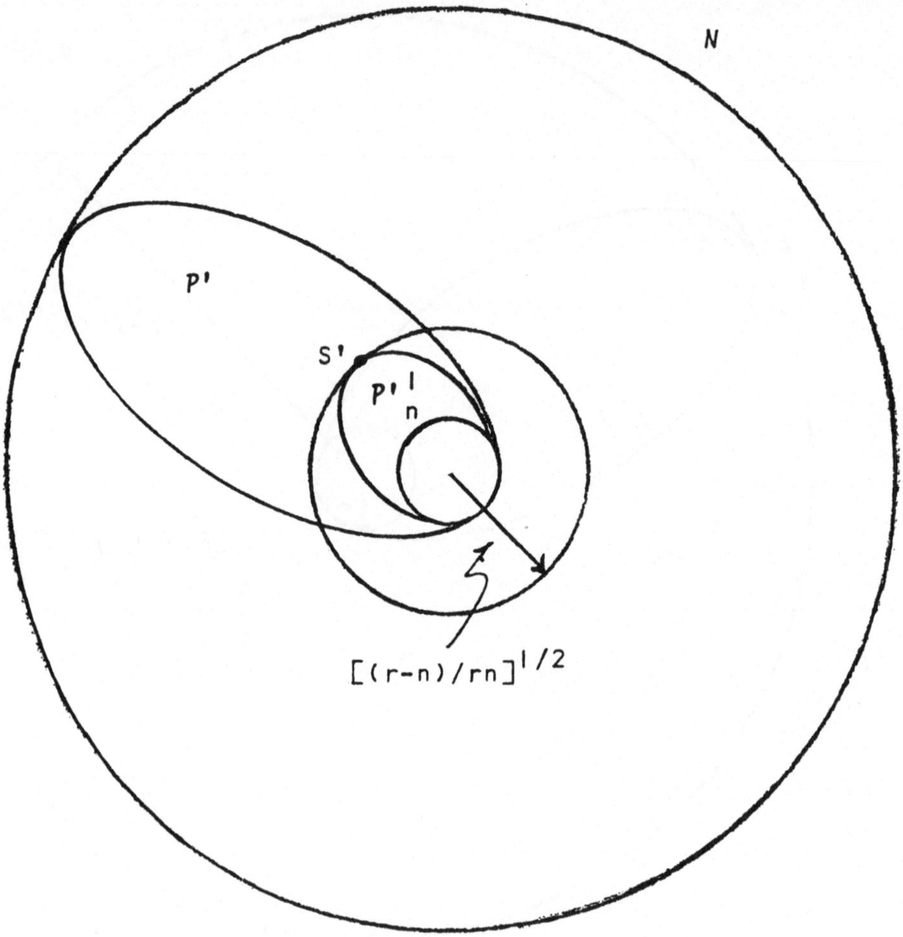

Figure 2. The subset P'^{1}_{n} in N.

The set of n-representable one-electron reduced density matrices is contained in a hypersphere of smaller diameter. The point of intersection S' corresponds to a Slater one-matrix.

References - Chapter II

Coleman, A. J. (1963), Rev. Mod. Phys. 35, 668.

Harriman, J. E. (1978), (a) Phys. Rev. A 17, 1249; (b) ibid., 1257; (c) Internat. J. Quant. Chem. 15, 611.

Kummer, H. (1967), J. Math. Phys. 8, 2063.

Kutzelnigg, W. (1963), Z. Naturforsch. 18a, 1058.

__________. (1965), Z. Naturforsch. 20a, 168.

Levy, M. (1979), Proc. Nat'l. Acad. Sci. USA 76, 6062.

III. APPLICATIONS OF THE GEOMETRY OF DENSITY MATRICES TO DENSITY FUNCTIONAL THEORY

A. Introduction

We will be concerned in this chapter with the question of how to extract from a density, ρ, some of the information which the Hohenberg-Kohn (1964) theorem tell us it contains. Two problems will be addressed: formulation of the constraints placed on a density matrix by the requirement that it correspond to a given density, and the expansion of a given density in terms of the eigenfunctions of a one-electron reduced density matrix (possibly idempotent).

In his algorithmic formulation of the Hohenberg-Kohn functional, Levy (1979) introduces a variational calculation with a wavefunction constrained to correspond to a particular density. We will consider here an n-electron density matrix so constrained. For a fixed, finite basis set density matrices are replaced by finite-dimensional matrices in a series of vector spaces, one for each number of electrons. Use of the reducing basis described in the previous chapter allows us to readily see what constraints are placed on an n-matrix by the requirement that it reduce to a given 1-matrix. The remaining problem of determining the set of 1-matrices which correspond to a given density will be addressed in Section B.

Instead of starting with a fixed basis set, we might think of using the density itself to define an appropriate set of orbitals. The term "appropriate" refers to two aspects of density functional theory: n representability and v representability. Gilbert (1975) showed that any density is n representable. An alternative approach is given in

Section C, where a presciption is presented for the construction of an arbitrary number of continuous, orthonormal functions whose squares can be combined to make up the given density.

The v-representability question asks whether for some given positive density there is a local potential such that the ground state of the Hamiltonian with this potential has the given density. [Larsen (1975)] In Section D, the trivial affirmative answer is given for a single-particle density. As a preliminary investigation in the many-particle case, noninteracting Fermions in one dimension are considered. Conditions which the functions must satisfy are obtained, and it is shown that these conditions are not satisfied by the functions developed in Section C, at least for the case of two particles.

B. Relationship between Density and Density Matrix with a Finite Basis Set

In the spirit of the geometric treatment presented in the previous chapter, we now assume the problem to be formulated in terms of some fixed, finite basis set. We want to use the same basis for both density and density matrix, so we assume that the density can be written as

$$\rho(\underline{r}) = \sum_{j,k} \rho_{jk} \, \phi_j(\underline{r}) \, \phi_k(\underline{r}) \tag{1}$$

where the $\{\phi_i\}$ are the given basis functions, assumed to be orthonormal, and it has also been assumed for simplicity here that everything is real. In this section we will consider only densities that can be so expressed. We will also avoid complications introduced by spin by assuming a singlet state so that the distinction between the one-matrix and the one-electron charge density matrix is trivial. (We could also treat separate densities for different spins, etc.) We can expand such a spatial density matrix as

$$\gamma(\underline{r};\underline{r}') = \sum_{j,k} \gamma_{jk} \, \phi_j(\underline{r}) \, \phi_k(\underline{r}') \tag{2}$$

and we are interested in those density matrices for which

$$\gamma(\underline{r};\underline{r}) = \rho(\underline{r}) \, . \tag{3}$$

We therefore seek a relationship between $\underset{\sim}{\rho} = (\rho_{jk})$ and $\underset{\sim}{\gamma} = (\gamma_{jk})$.

Once $\underset{\sim}{\gamma}$ or a set of possible $\underset{\sim}{\gamma}$'s is determined, methods described in the previous chapter will allow us to determined constraints placed on the n-matrix or other reduced density matrices. They must lie in the intersection of the space of n-matrices (or n-representable reduced density matrices) with the preimage of $\underset{\sim}{\gamma}$. This preimage space consists of matrices with some components which are completely determined by $\underset{\sim}{\gamma}$ and other components which are completely free.

From Eqs. (1)-(3) we see that γ_{jk} will be determined by ρ if we can uniquely isolate the coefficient ρ_{jk} in the expansion of ρ. This will be possible if and only if all the products $\phi_j(\underline{r})\phi_k(\underline{r})$ are linearly independent. A little consideration will show that most such products are independent. A typical basis function includes the product of a polynomial and an exponential. The exponential factors alone, in different products, will be different unless it happens that a sum of two orbital exponents is equal to a sum of two other orbital exponents. Exact linear dependence will thus be a rare accident. A more important practical consideration is the possibility of near linear dependence. This will surely occur if the basis set is sufficiently large.

Suppose that there are m basis functions ϕ_j. There are then potentially $m(m+1)/2$ distinct products $\phi_j\phi_k$. We can think of orthogonalizing them by the Schmidt orthogonalization procedure, and after we have M orthonormal combinations of the products it may be numerically infeasible to construct further orthonormal functions. The effective dimension of the product space is then $M < m(m+1)/2$.

Let $\{\Phi_J, 1 \leq J \leq M\}$ be the orthonormal functions spanning the product space. Then

$$\rho(\underline{r}) \cong \sum_{J=1}^{M} \rho_J \, \Phi_J(\underline{r}) \tag{4}$$

and

$$\phi_j(\underline{r})\phi_k(\underline{r}) \cong \sum_{J=1}^{M} C_J^{(jk)} \, \Phi_J(\underline{r}) \ , \tag{5}$$

where the equality may not be exact but can be considered as an equality within the numerical precision of the calculation. The coefficients ρ_J can be uniquely extracted from $\rho(\underline{r})$, and the $C_J^{(jk)}$ are known from the orthogonalization process. We then have

$$\sum_{j,k} C_J^{(jk)} \, \gamma_{jk} = \rho_J \qquad 1 \leq J \leq M \tag{6}$$

which fixes some of the γ_{jk} in terms of the others, and the latter can be considered as independent. It should also be noted that $\underset{\sim}{\gamma}$ must be subject to n-representability constraints. The point of interest here is that unless the basis set is quite large we must expect M to be nearly m(m+1)/2 so there are few independent parameters in $\underset{\sim}{\gamma}$. If $M = m(m+1)/2$, $\{C_j^{(jk)}\} = \underset{\sim}{C}$ is invertable and $\underset{\sim}{\gamma}$ is completely determined by $\underset{\sim}{\rho}$.

C. Expansion of a Given Density in Terms of Orthonormal Orbitals

It is commonly assumed in density functional theory that any density can be expressed as

$$\rho(\underline{r}) = \frac{1}{n} \sum_{j=1}^{n} |\phi_j(\underline{r})|^2 \qquad (7)$$

here the orbitals are orthonormal and we assume here that

$$\int \rho(\underline{r})d\underline{r} = 1 . \qquad (8)$$

We normalize ρ to 1 to emphasize that n is essentially arbitrary: if ρ is normalized to n then $\rho' = (n'/n)\rho$ is a perfectly good density normalized to n'. Gilbert (1975) has shown that Eqs. (7) and (8) can always be satisfied, and that and density is not only n representable but also pure state n representable by a single determinant. His construction is essentially to divde space into n distinct regions and take ϕ_j to be $\rho^{1/2}$ in the j-th region and zero elsewhere. Such functions are clearly orthogonal and can be made to be normalized if the sizes of the regions are appropriately chosen [for ρ normalized to n]. Such functions will be discontinuous and have discontinuous derivatives at the region boundaries, and although it seems plausible that these defects could be corrected, an explicit procedure for doing so has not been presented.

An alternative way of construction orthonormal functions satisfying Eq. (7) is presented here [Harriman (1981)]. Any number of orbitals can be constructed for a given density, and if desired ρ can be expressed in terms of more than n functions with occupation numbers less than 1/n. The functions are not unique, and it will be shown in the next section that they lack certain desirable properties.

We consider a one-dimensional case first. An obvious approach is to express each orbital as the product of an amplitude, equal to the square root of the density, and a phase factor, with the phases chosen to provide orthonormality. The problem is to find an appropriate choice of the phase functions. For a density $\rho(x)$ with $-\infty < x < \infty$, we let

$$f(x) = 2\pi \int_{-\infty}^{x} \rho(y)dy \qquad (9)$$

and take

$$\phi_k(x) = p(x) \exp[ik\, f(x)] \qquad (10)$$

for any integer k, where

$$p(x) = [\rho(x)]^{1/2} . \qquad (11)$$

The phase function f is monotone increasing with

$$f(-\infty) = 0 < f(x) < f(\infty) = 2\pi \tag{12}$$

and

$$\frac{df}{dx} = 2\pi\, \rho(x) \,. \tag{13}$$

It is obvious that

$$|\phi_k|^2 = [p(x)]^2 = \rho(x) \tag{14}$$

and

$$\begin{aligned}\int_{-\infty}^{\infty} \phi_k^*(x)\phi_\ell(x)dx &= \int_{-\infty}^{\infty} \rho(x)\, \exp[i(\ell-k)\; f(x)]dx \\ &= \int_{-\infty}^{\infty} \exp[i(\ell-k)f(x)]\, \frac{1}{2\pi}\frac{df}{dx}\, dx \\ &= \frac{1}{2\pi}\int_0^{2\pi} \exp[i(\ell-k)f]df = \delta_{k\ell} \,. \end{aligned} \tag{15}$$

The functions $\{\phi_k\}$ are orthonormal, smooth and continuous (assuming ρ is) and ρ can be expressed as

$$\rho(x) = \sum_k \lambda_k\, \phi_k(x)\, \phi_k^*(x) \tag{16}$$

for any set of λ_k's summing to 1. If ρ is to correspond to an n-representable density matrix

$$D^{(1)}(x;x') = \sum_k \lambda_k \, \phi_k(x) \, \phi_k^*(x') \tag{17}$$

it is necessary that $0 \leq \lambda_k \leq 1/n$. A particular possible choice of the λ's is to take n of them equal to 1/n and the rest zero, in which case $\gamma = nD^{(1)}$ will be idempotent.

Real functions in terms of which ρ can be expanded are also possible. Let

$$\psi_0(x) = p(x)$$

$$\left.\begin{aligned} \psi_{2j-1}(x) &= 2^{1/2}\, p(x) \sin[j\, f(x)] \\ \psi_{2j}(x) &= 2^{1/2}\, p(x) \cos[j\, f(x)] \end{aligned}\right\} j > 0 \, . \tag{18}$$

These functions can be shown, as in Eqs. (15), to be orthonormal and

$$[\psi_0(x)]^2 = \rho$$

$$[\psi_{2j-1}]^2 + [\psi_{2j}]^2 = \rho \, . \tag{19}$$

For them to be used to expand ρ or $D^{(1)}$ it is necessary that ψ_{2j-1} and ψ_{2j} have the same occupation number.

Extension to three dimensions is straightforward, but not unique. Let a,b,c be any set of orthogonal coordinates with $a_1 \leq a \leq a_2$, $b_1 \leq b \leq b_2$ and $c_1 \leq c \leq c_2$ and with volume element

$$d\tau = \alpha(a)\ \beta(b)\ \gamma(c)\ da\ db\ dc\ . \tag{20}$$

Then, for example,

$$\int_{a_1}^{a_2} da\ \alpha(a) \int_{b_1}^{b_2} db\ \beta(b) \int_{c_1}^{c_2} dc\ \gamma(c)\ \rho(a,b,c) = 1 \tag{21}$$

and we define p essentially as before

$$p(a,b,c) = [\rho(a,b,c)]^{1/2}\ . \tag{22}$$

The phase function f must be a function of a single variable if it is to become a variable of integration as in Eq. (15). We take a, which could be any one of the three variables in whatever set we are using. Then let

$$f(a) = 2\pi \int_{a_1}^{a} da'\ \alpha(a') \int_{b_1}^{b_2} db\ \beta(b) \int_{c_1}^{c_2} dc\ \gamma(c)\ \rho(a',b,c) \tag{23}$$

so that as before $0 \leqslant f \leqslant 2\pi$, and

$$\frac{df}{da} = \bar{\rho}(\alpha) = \int_{b_1}^{b_2} db\ \beta(b) \int_{c_1}^{c_2} dc\ \gamma(c)\ \rho(a,b,c)\ . \tag{24}$$

The orbitals are then given by equations analogous to Eq. (10) [or Eqs. (18)], but with x replaced by a,b,c in ϕ and p, and by a only in f. The remainder of the previous discussion generalizes in a straightforward way.

In a one-center problem, for example, it would be natural to take $a,b,c = r,\theta,\phi$ and $\bar{\rho}(r)$ is the spherical average of $\rho(\underline{r})$. It would be equally possible, however, to take $a = \theta$ or ϕ, and different functions would result. If $\rho(\underline{r})$ is in fact spherical, a contribution corresponding to any $\phi_k(\underline{r})$ can be further divided into contributions from $\chi_{k\ell m} = \phi_k(r)\, Y^m(\theta,\phi)$. Since

$$\sum_{m=-\ell}^{\ell} Y^m(\theta,\phi)[Y^m(\theta,\phi)]^* = \frac{2\ell+1}{4\pi} , \qquad (25)$$

an expansion of ρ in terms of the $\chi_{k\ell m}$ will be possible if coefficients are appropriately adjusted and all those for a given k and ℓ are equal, independent of m. It is clear that in the three-dimensional case there is much flexibility in the choice of functions as well as in the occupation numbers.

D. The Relationship between a Density and a Local Potential

In the case of a single particle there is a simple relationship between a density and a local potential. Given any $\rho(\underline{r}) \geqslant 0$, let

$$\psi(\underline{r}) = [\rho(\underline{r})]^{1/2} . \qquad (26)$$

The Schrödinger equation (in reduced units)

$$-\tfrac{1}{2}\nabla^2\psi + v(\underline{r})\psi = E\,\psi \qquad (27)$$

will be satisfied by ψ provided

$$v(\underline{r}) = \frac{\nabla^2\psi(\underline{r})}{2\psi(\underline{r})} + E \,. \qquad (28)$$

The zero point of the energy scale is arbitrary and any shift will affect v and E in the same way.

To complete the proof that $v(\underline{r})$ is the potential we want, we must show that ψ is the ground state solution for Eq. (27). Any expectation value is a weighted average of eigenvalues and we can conclude that ψ is the ground state if, for an arbitrary (normalized, well-behaved) function $\phi(\underline{r})$

$$\tilde{E} - E = \int \phi(\underline{r})[-\tfrac{1}{2}\nabla^2 + v(\underline{r})]\, \phi(\underline{r})\, d\underline{r} \geq 0 \qquad (29)$$

with equality only if $\phi = \psi$ for a nondegenerate ground state. (Again, for simplicity, we assume ϕ real. Complex ϕ could be treated as well, but the equations become somewhat messier.) In requiring ϕ to be "well behaved" we require it to vanish at infinity, as we assume ψ to do. We then substitute $v(\underline{r})$ from Eq. (28) and integrate by parts to transfer one $\underline{\nabla}$ to the other ϕ to obtain

$$2(\tilde{E}-E) = \int (\underline{\nabla}\phi)^2\, d\underline{r} + \int \frac{\phi^2(\nabla^2\psi)}{\psi}\, d\underline{r} \,. \qquad (30)$$

Similarly,

$$\int \left(\frac{\phi^2}{\psi}\right) \nabla^2\psi \, d\underline{r} = -\int (\underline{\nabla}\psi)\cdot\left[\underline{\nabla}\left(\frac{\phi^2}{\psi}\right)\right] d\underline{r}$$

$$= -\int \underline{\nabla}\psi\cdot\left[\frac{2\phi\underline{\nabla}\phi}{\psi} - \frac{\phi^2}{\psi^2}\underline{\nabla}\psi\right] dt$$

$$= \int \left[\left(\frac{\phi\underline{\nabla}\psi}{\psi}\right)^2 - 2\underline{\nabla}\phi \cdot \frac{\phi\underline{\nabla}\psi}{\psi}\right] \tag{31}$$

so that

$$2(\widetilde{E}-E) = \int \left[(\nabla\phi)^2 - 2\underline{\nabla}\phi \cdot \frac{\phi\underline{\nabla}\psi}{\psi} + \left(\phi\,\frac{\underline{\nabla}\psi}{\psi}\right)^2\right] d\tau$$

$$= \int \left[\underline{\nabla}\phi - \left(\frac{\phi}{\psi}\,\underline{\nabla}\psi\right)\right]^2 d\tau \geqslant 0 \tag{32}$$

with equality if and only if $\phi = \psi$.

The multi-particle case is much more difficult to treat, and as a preliminary investigation we will consider the case of noninteracting Fermions in one dimension. In such a case the ground state density is

$$\rho(x) = \frac{1}{n}\sum_{j=1}^{n} |\psi_j(x)|^2 \tag{33}$$

where $\psi_1,\ldots,\psi_n$ are the n lowest-energy solutions of a Schrödinger equation

$$-\tfrac{1}{2}\psi_j'' + v(x)\psi_j = e_j\psi_j \, . \tag{34}$$

We will see that, while it is easy to find orthonormal functions satisfying Eq. (33), the requirement that they all satisfy the same Schrödinger equation is not easily met.

Suppose that n functions satisfying Eqs. (33) and (34) exist. For any ψ_j, Eq. (34) can be solved to give v(x), as in Eq. (28). If two such equations, obtained from different functions, are subtracted, v(x) is eliminated and the result, multiplied by 2, can be written

$$\frac{\psi_j''}{\psi_j} - \frac{\psi_k''}{\psi_k} = 2(e_k - e_j) \; . \tag{35}$$

We note that by multiplying this equation by $\psi_j\psi_k$ and integrating we get

$$\begin{aligned} 2(e_k - e_j) \int \psi_j\psi_k \, dx &= \int (\psi_k\psi_j'' - \psi_j\psi_k'') \, dx \\ &= \int (\psi_k'\psi_j' - \psi_j'\psi_k') \, dx \\ &= 0 \end{aligned} \tag{36}$$

so that if $e_j \neq e_k$, ψ_j and ψ_k are orthogonal, a well-known result.

Let us now further specialize to the case $n = 2$. With $p = \rho^{1/2}$ as before and $\psi_i = p\phi_i$, $i = 1,2$

$$\phi_1^2 + \phi_2^2 = 1 \tag{37}$$

and Eq. (35) becomes

$$2\,\frac{p'}{p}\left(\frac{\phi_1'}{\phi_1}-\frac{\phi_2'}{\phi_2}\right)+\frac{\phi_1''}{\phi_1}-\frac{\phi_2''}{\phi_2} = \lambda \tag{38}$$

where $\lambda = 2(e_2 - e_1)$. We note also that

$$\frac{2p'}{p} = \frac{2pp'}{p} = \frac{\rho'}{\rho} = \frac{d}{dx}\,\ell n\,\rho \tag{39}$$

The constraint required by Eq. (37) can be satisfied in two ways. If we take

$$\phi_1 = \phi\,, \qquad \phi_2 = (1-\phi^2)^{1/2} \tag{40}$$

then Eq. (48) can be written

$$\frac{\left(\frac{\rho'}{\rho}\right)\left(\frac{\phi'}{\phi}\right)}{1-\phi^2} + \frac{\left(\frac{\phi'}{\phi}\right)}{(1-\phi^2)^2} = \lambda \tag{41}$$

while if we take

$$\phi_1 = \cos\alpha\,, \qquad \phi_2 = \sin\alpha \tag{42}$$

it is equivalent to

$$\alpha'' + \frac{\rho'}{\rho}\alpha' = \frac{\lambda}{2}\sin 2\alpha\,. \tag{43}$$

Neither of the nonlinear equations (41) or (43) is readily solved, especially for general ρ'/ρ .

If we take α to correspond to the f of Eq. (9), $\alpha' = 2\pi\rho$ and $\alpha'' = 2\pi\rho'$ so Eq. (43) is equivalent to

$$2\pi(\rho' + \frac{\rho'}{\rho}\rho) = \frac{\lambda}{2}\sin 2\alpha \tag{44}$$

or

$$2\alpha = \arcsin\left(\frac{8\pi}{\lambda}\rho'\right) \tag{45}$$

and differentiation gives

$$2\rho = \frac{\rho''}{[1 - (\frac{8\pi}{\lambda}\rho')^2]^{1/2}} . \tag{46}$$

This equation will surely not be satisfied by general densities ρ, so we can conclude that the functions constructed in Section C are not the physically meaningful functions which satisfy Eq. (35).

E. Conclusions

The relationship between a density and a density matrix can be considered in two ways. If both are assumed to be expressed in terms of a given, finite basis set then the density almost completely determines the density matrix, but there may be some freedom in the density matrix to the extent that products of basis functions are linearly dependent. The density can be used to define appropriate one-electron functions in terms of which it and the density matrix can be expressed. Many choices of functions and their occupation numbers are consistent with a given density. Functions of the form defined are unlikely to be eigenfunctions of a single Hamiltonian with a local potential, however. Even in the case of noninteracting Fermions in one dimension, functions which lead to the given density and are also solutions of the same Schrödinger equation must satisfy a density-dependent nonlinear differential equation. One is led to speculate that most densities are not v representable.

References - Chapter III

Gilbert, T. L. (1975), Phys. Rev. B 12, 2111.

Harriman, J. E. (1981), Phys. Rev. A (to be published).

Hohenberg, P. and Kohn, W. (1964), Phys. Rev. 136, B864.

Larson, E. G. (1975), Informal remarks at the Boulder Theoretical Chemistry Conference.

Levy, M. (1979), Proc. Nat'l. Acad. Sci. USA 76, 6062.

PROPERTIES OF ONE-MATRIX ENERGY FUNCTIONALS

Robert A. Donnelly
Auburn University

CONTENTS:

Supported in part by Grant ACS-PRF 12517-G6

I. INTRODUCTION

The Hohenberg-Kohn (1) theorem asserts that the energy E(N,v) of an atomic or molecular ground state is a well-defined and bounded functional of the electron density. In fact, assuming the ground state to be non-degenerate for simplicity, the density ρ and external potential v are in one-to-one correspondence with the wavefunction

(1) $$v \Leftrightarrow \rho \Leftrightarrow \Psi$$

From Eq. (1) it follows that the density determines the electronic energy via the functional of Hohenberg and Kohn

(2) $$E_v^{HK} \equiv \int d\tau\, \rho(\vec{r})\, v(\vec{r}) + F$$

The first term in Eq. (2) is the electron-nuclear attraction, the second contains all remaining constributions to the electronic energy. Though its precise form remains at present unknown this functional obeys a variational principle

(3) $$E(N,v) = \min_{\rho' \in P} E_v^{HK}[\rho']$$

provided the set P is restricted to those densities which are normalized to the correct number of electrons. P must also be restricted to contain densities which are simultaneously N- and V-representable (2) .

Several authors have demonstrated the existence of an energy functional based on the one-matrix. Gilbert came to this conclusion in extending the Hohenberg Kohn theorem to systems with a non-local external potential (3) . Berrondo and Goscinski (4) performed a Legendre transformation on the two-matrix functional, obtaining equations satisfied by a one-matrix functional. Donnelly and Parr (5) discussed the one-matrix functional for a pure state and the commutation rules satisfied by the one-matrix of a stationary state (One of these rules was also obtained by Berrondo and Goscinski). Donnelly has shown at least one fundamental difference between energy functionals beased on one- and on two-matrices (6) . Levy (7) has given a lucid discussion of the density and density matrix energy functionals, and Valone (8) has begun an extension of the theory to excited states. Valone (9) has also considered the extension of the density matrix theory to systems described by ensemble-representable one-matrices, a topic also discussed by Gilbert (3) . Harriman has given a discussion (10) of the geometry of the set of N-representable one matrices, and there was recently given a discussion of the V-representability problem (11) . Interesting criticisms of the density functional theory have been given by Primas (12) and by Kryachko (13) .

In Section II we discuss the application of the variational principle to the energy functional. There we give the conditions on the occupation numbers and natural spinorbitals of an approximate ground-state one-matrix. The physical interpretation of these results follows upon identifying the negative of the electronegativity with the chemical potential (14) . Sanderson's principle (15) of electronegativity equalization then follows.

In Section III we discuss the one-matrix theory and its relation to the equations of the multi-configuration self-consistent field method. This comparison is motivated by the realization that both theories reduce to the same common first approximation, the Hartree-Fock method. This comparison is likewise stimulated by the similarity of the quantities ionization potential and chemical potential.

In Section IV we introduce an approximate form of the one-matrix theory, which includes the Hartree-Fock-Slater (16) theory as a special case. This functional is interesting, in that it satisfies the commutation rules of Section II, but does not exhibit the degeneracy of the eigenvalue spectrum found in Section II for the exact functional. Concluding remarks follow in Section V.

II. AN EXACT ONE-MATRIX ENERGY FUNCTIONAL

In this Section we specify and interpret the conditions on the occupation numbers and natural spinorbitals of a finite-rank approximation to the exact ground state one-matrix. Our treatment is restricted to the special case of an atom or molecule with a non-degenerate ground state, though the same equations apply to ensemble-representable one matrices (3) . Treatment of this state is especially simple, in view of the fact that the necessary and sufficient conditions for both ensemble- and pure-state representability of one-matrix are known (17) . As pointed out by Gilbert, the situation presented by pure, degenerate ground states is more difficult, since necessary and sufficient conditions on N-representability in this case are unknown at present (3) .

Consider a system with N electrons, local potgntial v , and Hamiltonian H(N,v) . The ground state wavefunction obeys Schrödinger's equation

$$H\Psi = E(N,v)\Psi \ , \tag{4}$$

and the reduced density matrix of order $1 \leqq P \leqq N$ describing the electron distribution is defined in the

Löwdin (18) normalization by the formula

(5) $\Gamma^{(p)}(x_1, x_2, \cdots x_p | x_1', x_2' \cdots x_p') \equiv$

$$\binom{N}{p} \int \cdots \int dx_{p+1} \cdots dx_N \, \Psi(x_1, x_2 \cdots x_p \cdots x_N) \Psi^*(x_1', x_2' \cdots x_p', x_{p+1} \cdots x_N)$$

The coordinate x is a vector, $x = \vec{r}s$, representing the space ($\vec{r}$) and spin (s) coordinate of each electron, and an N-representable density matrix (one describing a fermion system) must be an antisymmetric positive semi-definite hermitian form (17) . An N-representable density matrix must further be of trace-class; it has a spectrum of discrete eigenvalues which in the special case $p = 1$ must lie on the interval $[0 - 1]$.

II. A Variational Principle

Let us begin by defining the space KER_1 of trace-class bilinear forms on Hilbert space, and the association G mapping KER onto the set Γ of positive semi-definite one matrices which are N- and V-representable by pure states

(6) $G : Ker_1 \rightarrow \Gamma \ni \gamma$

Choosing a complete basis $\{\Psi\} = \mathcal{K}$ for KER_1 the one-matrix is representable by the bilinear form in Eq. (7)

(7) $$\gamma(x,x') = K(x)\,\underline{\underline{\gamma}}\,K^{+}(x)$$

For orthonormal K the norm of $\underline{\underline{\gamma}}$ is its trace

(8) $$n[\gamma] = tr\,\underline{\underline{\gamma}} = \sum_p \langle \psi_p | \underline{\underline{\gamma}} | \psi_p \rangle$$

which is finite for any choice of basis on Ker_1 . The natural spinorbitals and occupation numbers of K are related by the eigenvalue equation below

(9) $$\int d\xi\, \gamma(x,\xi)\,\chi_p(\xi) = n_p\,\chi_p(x)$$

Thus in the natural spinorbital representation the one-matrix takes the diagonal form

(10) $$\gamma(x,x') = \sum_p \chi_p(x)\cos^2\theta_p\,\chi_p^*(x')\;; \qquad 0 \le \theta_p \le \frac{\pi}{2}$$

Here we use the parameterization given by Gilbert (3) . It is especially convenient since it enforces the positivity of the occupation numbers while restricting their range to the physically admissable interval [0 - 1] . In what follows we consider approximations to the exact $\underline{\underline{\gamma}}$ which are themselves of finite rank, further restricting the discussion to the representation in which $\underline{\underline{\gamma}}$ is diagonal (A more restricted treatment which is representation independent has also been given (5)) .

The minimum energy can be found in a finite dimensional setting by application of an optimization procedure similar to that used by Löwdin (18) in his exposition of MCSCF theory. To see this choose a finite integer $M > N$ (N is the number of electrons) and denote by $KER^{(\beta)}$ ($\beta = 1,2,3,\ldots$) a particular rank-M subdomain of the space KER_1 defined previously. Each $KER^{(\beta)} \subset KER_1$ is spanned by a set of basis vectors indexed by β which we denote by $K^{(\beta)} \subset K$. The set of allowable one matrices of finite rank is denoted by $\Gamma^{(\beta)} \subset \Gamma$. The approximate one-matrix can thus be written in the form

$$(11) \qquad \gamma^{(\beta)} = K^{(\beta)} \underline{\underline{\gamma}} \, K^{(\beta)+} .$$

This shows clearly the dependence of this function on the occupation numbers and the particular subspace of KER_1 which has been chosen.

Denote the one-matrix energy functional by the symbol $E^{M}_{\upsilon}[\gamma]$ in which the integer M indicates that the functional is applied to functions of finite rank (M is not a variable in the optimization scheme discussed below). The optimum approximate ground state energy then satisfies the inequality $E^M(N,v) \leq E(N,v)$ provided $\gamma^{(\beta)}$ satisfies the normalization constraint $\mathcal{N}[\gamma^{(\beta)}] = N$. The optimum approximate ground state energy is thus found by implementation of the optimization scheme in Eq. (12) below

$$(12) \quad E^M(N,v) = \min_{K^{(\beta)} \subset K} \left\{ \min_{\gamma^{(\beta)} \subset \Gamma^{(\beta)}} E_v^M[\gamma^{(\beta)}] \right\} ;$$

to simplify the notation we henceforth dispense with the superscript β .

Associate the Lagrange multiplier μ with the normalization constraint of Eq. (8) and the matrix of Lagrange multipliers $\underline{\underline{\Lambda}} = \{\lambda_{ij}\}$ with a set of orthonormality constraints on the spinorbitals

$$(13) \quad S_{ij} = \langle \chi_i | \chi_j \rangle = \delta_{ij}$$

The $\{\lambda_{ij}\}$ have the dimension of energy; we shall denote the eigenvalues by the special symbol $\{\varepsilon\}$, and define these as the orbital energies. Now define the auxilliary functional Ω by the formula

$$(14) \quad \Omega \equiv E_v[\gamma] - \mu \mathcal{N}[\gamma] - \mathrm{tr}\, \underline{\underline{\Lambda}}\, \underline{\underline{S}} \, .$$

A variation in Ω accompanying variation in the independent parameters $\{\theta\}$, the natural spinorbitals, and their conjugates takes the form

$$\delta\Omega = \sum_p \left\{ \frac{\partial E_0}{\partial \theta_p} + \mu \sin 2\theta_p \right\} \delta\theta_p + \tag{15}$$

$$\sum_p \left\{ \int d\xi \left[\frac{\delta E}{\delta \chi_p(\xi)} - \sum_q \chi_q^*(\xi) \lambda_{qp} \right] \delta \chi_p(\xi) \right.$$

$$+ \text{complex conjugate.}$$

A condition on a stationary one-matrix is that the first variation in Ω vanish, $\delta\Omega = 0$, from which one obtains the conditions on the optimum occupation numbers and spinorbitals. Introducing the variational derivative of the energy with respect to the one-matrix $F[\gamma, x', x]$ the various derivatives in this last equation can be directly related to $F[\gamma]$ itself by use of the functional chain rule. After some algebra, the partial derivative with respect to θ_p can be shown to be given by the formula

$$\frac{\partial \Omega}{\partial \theta_p} = \frac{\partial E_0}{\partial \theta_p} + \mu \sin 2\theta_p = -\sin 2\theta_p \,(F_{pp} - \mu) \tag{16}$$

where the matrix element F_{pp} is simply the derivative of the energy with respect to the occupation number n_p

(17)

$$F_{pp} = \iint dx\, dx'\, F[\gamma; x', x']\, \chi_p(x)\, \chi_p^*(x') = \frac{\partial E_0}{\partial n_p}$$

From the independence of the parameters (θ) there follows the conditions on the occupation numbers, namely

(18) $$\sin 2\theta_p \,(F_{pp} - \mu) = 0 \qquad \text{for } p = 1,2 \cdots M.$$

From this it follows that all partially-occupied orbitals belong to the degenerate eigenvalue μ of $\underline{\underline{F}}$. On the other hand, orbitals which are either completely empty or completely occupied belong to arbitrary eigenvalues ε of $\underline{\underline{F}}$. Let $\underline{\underline{\gamma}}$ be written in the form of the direct sum $\underline{\underline{\gamma}} = \underline{\underline{P}} \oplus \underline{\underline{n}}(\underline{\underline{I}} - \underline{\underline{P}})$ where $\underline{\underline{P}}$ is that part of g which has fully occupied or fully empty orbitals. Let the complement to $\underline{\underline{P}}$ be $(\underline{\underline{I}} - \underline{\underline{P}})$ and $\underline{\underline{n}}$ be the diagonal matrix of occupation numbers which satisfy the strict inequality $0 < n_p < 1$. Then one obtains the equation

(19) $$\underline{\underline{F}}\,\underline{\underline{\gamma}} = \underline{\underline{\varepsilon}}\,\underline{\underline{P}} + \mu \underline{\underline{n}}\,(\underline{\underline{I}} - \underline{\underline{P}})$$

In the Hartree Fock case, $\underline{\underline{I}} - \underline{\underline{P}}$ is null by definition, which leads to the well-known equation

$$\underline{\underline{F}}\,\underline{\underline{P}} = \underline{\underline{\varepsilon}}\,\underline{\underline{P}} \tag{20}$$

At the other extreme when all $0 < n_p < 1$ one obtains the result

$$\underline{\underline{F}}\,\underline{\underline{\gamma}} = \mu\,\underline{\underline{n}}\,(\underline{\underline{I}} - \underline{\underline{P}}) \tag{21}$$

These equations imply the commutation rule

$$[\underline{\underline{F}}, \underline{\underline{\gamma}}] = \underline{\underline{0}} \tag{22}$$

for a stationary one-matrix. This well-known condition on the Hartree-Fock one matrix was obtained by McWeeny (19) . We (5) , Gilbert (3) , and Berrondo and Goscinski (4) have shown that this commutation rule holds in the general case, where the one-matrix is not required to be idempotent.

The equation satisfied by the optimum spinorbitals is

$$n_p F \chi_p = \sum_q \chi_q \lambda_{qp} \tag{23}$$

which may also be written in the form

$$F\chi_p = \sum_q \chi_q \lambda_{qp} n_p^{-1} \tag{24}$$

for the occupied orbitals only. The matrices $\underline{\underline{F}}$, $\underline{\underline{\gamma}}$, and $\underline{\underline{\Lambda}}$ are hermitian for a stationary state; thus this equation leads to the further commutation rule

$$[\underline{\underline{\Lambda}}, \underline{\underline{\gamma}}] = \underline{\underline{0}} \; . \tag{25}$$

In other words, in the natural spinorbital representation for the one-matrix, the Lagrange multiplier matrix $\underline{\underline{\Lambda}}$ and $\underline{\underline{\gamma}}$ are simultaneously diagonal. It follows that the spin-orbitals satisfy an eigenvalue equation

$$F \chi_p = \mu_p \chi_p \qquad \text{for } p = 1, 2, \cdots, M \, . \tag{26}$$

This commuation rule relating the occupation numbers and orbital energies is a significant and unique characteristic of a theory in which the energy is obtained from the one-matrix. In fact it may ultimately play a central role in relating the present theory to existing many body theories of quantum chemistry and quantum statistical mechanics. Specifically, as the orbital energies and occupation numbers are simultaneously well-defined quantities, the one matrix may be written in the form of Eq. (27)

$$\gamma(x, x') = \sum_p \chi_p(x)\, \omega_p(\varepsilon_1 \cdots \varepsilon_M)\, \chi_p^*(x') \tag{27}$$

in which the weights are energy-dependent by Eq. (25) . A consistent interpretation of this equation in "many-body" terms would be provided by regarding the natural spin-orbitals as the single-particle states occupied in the ground state of an interacting system. A similar idea was expressed by Husimi (20) in his original work introducing the reduced one-matrix: The commutation rules in Eqs. (22) and (25) are precisely those which one would impose in order that an ensemble such as that of Eq. (27) be stationary.

II. B Electronegativity Equalization

The physical content of these equations has been interpreted by Parr and coworkers (14) and by Donnelly and Parr (3) by identification of the chemical potential with the negative of the electronegativity. The result of such an interpretation is a verification of Sanderson's Principle of electronegativity equalization, which may be stated as follows (15) :

> When two or more atoms initially different in electronegativity combine chemically their electronegativities become equalized in the molecule. The equalization of electronegativity occurs through the

adjustment of the polarities of the bonds which is pictured as resulting in a partial charge on each atom. That is electron loss causes increase, and electron gain decrease in electronegativity.

Consider the case in which to atoms A and B interact, forming the molecule AB. The electronegativity of atom A is $X_A = -\mu_A$, that of atom B is $X_B = -\mu_B$; for definiteness let $U_A < U_B$. The chemical potential of AB is μ_{AB}, the negative of the electronegativity of the molecule. Clearly, the chemical potential of the molecule is a single number which is characteristic for that molecule, while the chemical potentials of its precursors share this same property. Inasmuch as μ is the derivative of the energy with respect to electron number it follows that whenever μ_A and μ_B are different the energy of the system can be lowered by transfer of charge from one atom to another. Further, the initial direction of charge transfer is in the direction of lower chemical potential (higher electronegativity). A stationary molecular charge distribution clearly results when no further energy gain can be accomplished by electron redistribution, as is the case when the electronegativities of the atoms have become equalized in the molecule.

This view has added merit in view of recent attempts at definition of an orbital eletronegativity (21). The idea is to distinguish between core and valence orbitals on the basis of the tendency to acquire or donate electrons when atoms combine in molecule formation. One must admit that this idea has great intuitive appeal, but Eq. (21) shows that no such distinction can in fact be made. In particular let natural spinorbitals χ_i and χ_j belong respectively to atoms A and B above. Then

$$\mu_A = -\chi_A = \frac{\partial E_v}{\partial n_i} < \mu_B = -\chi_B = \frac{\partial E_v}{\partial n_j} \quad (28)$$

These inequalities show that the energy of the molecule AB can be lowered by transfer of charge from orbital j (on atom B) to orbital i (on orbital A) whenever the initial electronegativity of B is less than that of A. The driving force behind such transfer--the lowering of the energy of AB--disappears when the chemical potentials (electronegativities) become equal. Sanderson's principle is verified even when one considers the individuel orbitals in the system of interest.

III. COMPARISON OF THE ONE-MATRIX THEORY WITH THE EXTENDED HARTREE-FOCK EQUATIONS

In this Section we exhibit a distinction between the one-matrix theory discussed previously and a two-matrix theory which leads to the extended Hartree-Fock equations. The equations satisfied by the natural spinorbitals in the respective methods is remarkably similar and worthy of comment. Moreover, the extension of Koopmans' theorem proposed indepently by Smith and Day (22) and by Morell, Parr, and Levy (23) identifies eigenvalues of the variational derivative in the two-matrix theory as ionization potentials of the system of interest. Thus a comparison of the variational derivatives of these two operators (they act as "self-consistent" potentials) should serve to illuminate similarities and differences in the underlying energy functionals.

First, we write the variational equation for the orbitals in the two-matrix theory in a form which allows their comparison with the corresponding equation encountered in the one-matrix theory. We then show that the resulting operator is non-hermitian when one goes beyond the single determinant approximation. Thus the commutation rules given for the one-matrix theory are not satisfied by the corresponding quantities which appear in the multi-configuration SCF equations.

III. A The Functional $E_v[\Gamma^{(2)}]$.

Let the wavefunction for a system of N electrons and local potential v be represented in a finite basis $K^{(\beta)} = \{\psi\}$ of spinorbitals by the sum

$$\Psi = \sum_p \Phi_p C_p \tag{29}$$

in which the $\{C\}$ are complex numbers and the functions $\{\Phi\}$ are Slater determinants

$$\Phi_p \equiv det[\psi_{p1}(x_1)\psi_{p2}(x_2)\cdots\psi_{pN}(x_N)] \,. \tag{30}$$

Simultaneous determination of the coefficients $\{C\}$ and spinorbitals $\{\psi\}$ is one object of MC-SCF theory. The energy optimization scheme parallels that used in the previous Section. Let $C^{(\beta)}$ be a finite length coefficient vector and $K^{(\beta)}$ be a finite set of spinorbitals drawn from a complete set K. Then E (N,v), the optimum approximate ground state energy satisfies eq. (31) below

$$E^M(N,v) = \min_{K^{(\beta)}\subset K} \{ \min_{C^{(\beta)}\subset C} E_v[\Gamma^{(2)}] \} . \tag{31}$$

An auxilliary functional used in approximate solution of the Schrödinger equation may be defined as follows

$$\Omega \equiv E_v[\Gamma^{(2)}] - E^M(N,v)\sum_p |C_p|^2 - \mathrm{tr}\,\underline{\underline{\Lambda}}^L \underline{\underline{S}} \,. \tag{32}$$

Here $\Gamma^{(2)}$ is the two-matrix and the energy functional is defined by the formula

$$E_v[\Gamma^{(2)}] \equiv \iint dx\,dx'\, h(x,x')\,\gamma(x,x') + \iint dx_1 dx_2 \frac{\Gamma^{(2)}(x_1,x_2|x_1,x_2)}{r_{12}} \tag{33}$$

for the one-electron hamiltonian h(x,x')

$$h(x,x') \equiv \delta(x-x')\left[-\tfrac{1}{2}\nabla_{x'}^2 + v(x')\right] \tag{34}$$

and one-matrix γ . The Lagrange multipliers $\underline{\underline{\Lambda}}^L$ are introduced in order to constrain orthonormality of the basis:

$$S_{ij} = \langle \psi_i | \psi_j \rangle = \delta_{ij} \tag{35}$$

They form a hermitian matrix when optimization of the energy is complete. The Lagrange multiplier for the normalization is the ground state energy.

Variation of the spinorbitals in Eq. (33) while holding the down coefficients in the wavefunction fixed yields the

Euler equation satisfied by the optimum finite set. After some algebra (18) the result is

$$(36) \quad h \sum_q \psi_q \chi_{qp} + 2 \iint d\xi \, dx_2 \, \psi_q(\xi) \frac{\Gamma^{(2)}(\xi x_2 | x_1 x_2)}{r_{\xi 2}} = \sum_q \psi_q \lambda^L_{qp}$$

With $\underline{\underline{U}}$ the matrix which diagonalizes the one-matrix applied to the basis orbitals there results the equation satisfied by the in the form given first by Löwdin (18)

$$(37) \quad h \chi_p(x) n_p + 2 \iint d\xi \, dx_2 \, \chi_p(\xi) \frac{\Gamma^{(2)}(\xi x_2 | x_1 x_2)}{r_{\xi 2}} = U^{EHF} \chi_p(x),$$

$$U^{EHF} \chi_p(x) = \sum_q \chi_q(x) \lambda^L_{qp}$$

Equation (37) can be put in a form reminiscent of the orbital equation derived earlier for the one-matrix theory. Dividing through by $n_p \neq 0$, there results the equation

$$(38) \quad h \chi_p + \frac{2}{n_p} \iint d\xi \, dx_2 \, \chi_p(\xi) \frac{\Gamma^{(2)}(\xi x_2 | x_1 x_2)}{r_{\xi 2}}$$

$$= \sum_q \chi_q \lambda^L_{qp} n_p^{-1}$$

More compactly, define the operator V_{op} by the formula

$$(39) \quad V_{op} \chi_p(x) \equiv \frac{2}{n_p} \iint dx_2 \, d\xi \, \chi_p(\xi) \frac{\Gamma^{(2)}(\xi x_2 | x_1 x_2)}{r_{\xi 2}}$$

Now define the operator F_{op} by the formula

$$F_{op}\chi_p = h\chi_p + V_{op}\chi_p \tag{40}$$

There results the equation

$$F_{op}\chi_p = \sum \chi_q \lambda^L_{qp} n_p^{-1} \tag{41}$$

III. B Comparison of the Functionals $E_v[\gamma]$ and $E_v[\Gamma^{(21)}]$

The functionals $E_v[\gamma]$ and $E_v[\Gamma^{(21)}]$ can be seen to be different by comparison of the operators $F[\gamma]$ and F_{op} themselves. That is, in a matrix representation, the former is hermitian for the ground state one-matrix; thus the matrices $\underline{\underline{\Lambda}}$ and $\underline{\underline{\gamma}}$ commute, and the orbital energies and occupation numbers in the one-matrix theory are simultaneously well defined. The matrix $\underline{\underline{F}}_{op}$ is, however, non-hermitian for the ground state, since the operator which it represents is different for each different natural spin-orbital. Thus the corresponding quantities which arise in the two-matrix theory do not commute

$$[\underline{\underline{\Lambda}}^L, \underline{\underline{\gamma}}] \neq \underline{\underline{0}} . \tag{42}$$

The one- and two-matrix theories are different because they lead to different commutation rules. It follows the potentials which are central to the one- and two-matrix energy functionals are fundamentally different.

In light of this result, and in order to provide the material necessary for the next Section, we digress briefly to sketch the Extended Koopmans' Theorem which identifies the eigenvalues and eigenfunctions of the extended Hartree-Fock potential, defined by Eq. (37) , It follows as the result of three elementary theorems in the exposition of Morell, Parr, and Levy (23) which we present without proof. Theorem I allows a factorization of the total wave-function into sums of products of natural spinorbitals and complementary functions which are normalized to the occupation numbers of the one-matrix. Theorem II identifies the Lagrange multipliers appearing in Löwdin's equations. The Extended Koopmans' Theorem then follows by application of a variational principle.

III. C Extended Koopmans' Theorem

By means of the Extended Koopmans' theorem (22,23) it is possible to extract ionization potentials from a wave-function of multi-configuration type (The single-configuration results constitute ordinary Koopmans' theorem).

Theorem I (Carlson and Keller (24)) . The N-electron wavefunction Ψ may be written as a sum of products of natural spinorbitals and complementary functions $\{\Phi^{(N-1)}\}$ defined by the formulas

(43) $$\Phi_m^{(N-1)}(x_1 \cdots x_{N-1}) \equiv \sqrt{N} \int dx_N \, \chi_m(x_N) \Psi(x_1 \cdots x_N),$$

(44) $$\Psi(x_1 \cdots x_N) \equiv \frac{1}{\sqrt{N}} \sum_m \Phi_m^{(N-1)}(x_1 \cdots x_{N-1}) \chi_m(x_N).$$

The $\{\Phi^{(N-1)}\}$ are antisymmetric for fermions, orthogonal, and normalized to the occupation numbers of the one-matrix

(45) $$\langle \Phi_p^{(N-1)} | \Phi_q^{(N-1)} \rangle = \sqrt{n_p n_q} \, \delta_{pq} .$$

THEOREM II (Morell, Parr, Levy (23)) Consider an N-electron system with Hamiltonian $H(N,v)$ in eigenstate Ψ with energy $E(N,v)$. Let $\{\chi\}$ be the natural spinorbitals asscoiated with Ψ , with occupation numbers $\{n\}$. Let $\{\Phi^{(N-1)}\}$ be the functions complementary to the $\{\chi\}$ and let Γ be the two-matrix associated with Ψ . Then the quantities λ^L_{pq} , elements of a hermitian matrix

(46) $$\lambda^L_{pq} \equiv \langle \chi_p | U^{EHF} \chi_q \rangle$$

satisfy the identity

(47) $$\lambda^L_{pq} = \langle \Phi_p^{(N-1)} | E(N,v) - H(N-1,v) | \Phi_q^{(N-1)} \rangle.$$

The operator H(N-1,v) is the Hamiltonian for a system of (N-1) electrons and potential v , E(N,v) is a constant. The $\{\Phi^{(N-1)}\}$ are normalized according to Eq. (45) .

Theorem III (Morell, Parr, Levy (23)) define the matrix $\underline{\underline{\gamma}}$ by the formula

(48) $$\gamma_{pq} \equiv \lambda^L_{pq} / \sqrt{n_p n_q} \ .$$

Then the stationary states of the Hamiltonian E(N,v) - H(N-1,v) in the basis satisfy the system

(49) $$(\underline{\underline{\gamma}} - \gamma \underline{\underline{I}}) \underline{\underline{V}} = \underline{\underline{0}}$$

(50) $$\Psi_\alpha^{(N-1)} = \sum_p \Phi_p^{(N-1)} V_{p\alpha}$$

The roots of the determinantal equation

(51) $$\det [v_{pq} - v\, \delta_{pq}] = 0$$

bound ionization potentials of the N-electron parent system

(52) $$v^{(\alpha)} \geq - I^{(\alpha)} \equiv E(N,v) - E_{\alpha}(N-1,v).$$

The eigenfunctions are defined as natural transition orbitals. They approximate states (α) of the (N-1)-electron system obtained by removal of a single electron from Ψ .

An equivalent prescription for extraction of ionization potentials was given by Smith and Day (22) , who noted that the one-matrix plays the role of a metric with respect to which the Extended Hartree-Fock potential of Eq. (37) may be diagonalized. They thus define ionization potentials as roots of the determinantal equation

(53) $$\det [\lambda^{L}_{pq} - v\, \gamma_{pq}] = 0$$

which is completely equivalent to that used by Morell, Parr, and Levy. The natural transition orbitals thus obey the generalized eigenvalue problem

(54) $$U^{EHF}\chi_p(x) = \nu \int d\xi\, \gamma(x,\xi)\,\chi_p(\xi)\,.$$

It is the latter, rather than the natural spinorbitals, which play the fundamental role in ionization and attachment processes for atoms and molecules.

IV. AN APPROXIMATE ONE MATRIX ENERGY FUNCTIONAL

The operator F_{op} defined above is non-hermitian. Thus the MC-SCF theory cannot be put in a form in which the orbital energies and occupation numbers are simultaneously well-defined quantities. But in view of the theory in Section II it makes sense to search for hermitian variants of F_{op} , for this would lead immediately to an energy functional which obeys the commutation rules for the exact one. In this Section we approximate F_{op} by a procedure due to Löwdin (18) following work by Slater (16) . The result is an approximate one-matrix energy functional which reduces to the Hartree-Fock-Slater equations for an idempotent one-matrix. It is significant that Löwdin derived his result in a publication predating the Hohenberg-Kohn theorem by several years. Thus the existence of the latter theorem permits re-interpretation of the Löwdin-Slater theory in a wholly new context (25) .

The aim here is the construction of a hermitian form of the correct potential V_{op} . As this is different for each orbital, we thus consider an averaging procedure which results in an operator which is the same for all orbitals. To that end, consider the function W defined below

$$(55) \quad W = \sum_{p} \omega_p \, | V_{op} \chi_p - V_{av} \chi_p |^2$$

in which V_{av} is the "best local approximation" (18) to the correct potential V_{op} , and $\{\omega\}$ is a set of arbitrary weights. We shall define this best approximation by the condition that W vanish for a fixed set of weights. Now the choice of the weights is completely arbitrary, but one "obvious" procedure would be to take them as the occupation numbers of the one-matrix. That is, as the occupation numbers are strictly positive, W is a convex function, with its minimum, W = 0 , attained when for each value of the index p the following relation holds

$$(56) \quad V_{av} \chi_p(x_1) = V_{op} \chi_p(x_1) = \frac{2}{n_p} \iint d\xi \, dx_2 \, \chi_p(\xi) \frac{\Gamma^{(2)}(\xi x_2 | x_1 x_2)}{r_{\xi 2}}$$

The result is that the optimization problem for V_{av} has been "solved" by a simple choice of the weights themselves! Construction of V_{av} now proceeds by multiplication of Eq. (56) by $n_p \chi_p^*$ followed by summation. The result is

$$(57)\quad V_{av}(x) = \int dx_2\, \gamma^{-1}(x,x)\, \frac{\Gamma^{(2)}(x_1x_2|x_1x_2)}{r_{12}}$$

were use has been made of the identity (18)

$$(58)\quad \sum_p \int d\xi\, \chi_p^*(x)\chi_p(\xi)\,\Gamma^{(2)}(\xi x_2|x_1x_2) = \Gamma^{(2)}(x x_2|x_1 x_2) .$$

This admittedly ad hoc procedure results in a hermitian operator with which to replace its exact counterpart in an orbital equation. But continuing one further step, it is mathematically responsible to require the weights to be bounded in order that no single term dominate W . As normalization is a linear condition on the sum of the occupation numbers we shall simply require that they sum to the number of electrons in the system of interest. This condition is imposed by parameterization of N by the chemical potential.

Upon replacement of the exact potential in Eq. (39) by its best local approximation there results the equation

$$(59)\quad h\chi_p + V_{av}\chi_p = \sum_q \chi_q \lambda_{qp}^{LS} n_p^{-1} ,$$

which may be written more compactly in the form

$$(60)\quad F^{LS}\chi_p = \sum_q \chi_q \lambda_{qp}^{LS} n_p^{-1} .$$

The approximate self-consistent potential defined here is just that introduced into Hartree-Fock theory by Slater (16) . Löwdin generalized this to the case where the one-matrix is not required to be idempotent. We recognize the approximate nature of this equation by renaming the multipliers. It easily follows that the energy in this approximation is fixed by the one-matrix as we now demonstrate.

IV. A An Approximate One-Matrix Energy Functional

With a matrix representation understood, the orbital equation of what we shall term the Löwdin-Slater energy functional can be written as follows

(61) $$\underline{\underline{F}}^{LS} = \underline{\underline{\Lambda}}^{LS}\, \underline{\underline{\gamma}}^{-1} .$$

It follows immediately from the hermiticity of $\underline{\underline{F}}^{LS}$ (see Eq. (59)) and of the matrices $\underline{\underline{\gamma}}$ and $\underline{\underline{\Lambda}}^{LS}$ that the matrices $\underline{\underline{\gamma}}$ and $\underline{\underline{\Lambda}}^{LS}$ commute

(62) $$[\, \underline{\underline{\Lambda}}^{LS} , \underline{\underline{\gamma}} \,] = \underline{\underline{0}}$$

Likewise it follows that the matrices $\underline{\underline{F}}^{LS}$ and $\underline{\underline{\gamma}}$ commute

(63) $$[\underline{\underline{F}}^{LS}, \underline{\underline{\gamma}}] = \underline{\underline{0}}$$

so that the Löwdin-Slater energy functional obeys the commutation rules established earlier for the exact one-matrix energy functional.

The energy in this approximation is given by the formula

(64) $$E^{LS} = \frac{1}{2} \Sigma (n_p h_{pp} + \varepsilon_p^{LS})$$

in which h_{pp} is a matrix element of the one-electron operator defined in Eq. (34) and ε^{LS} is an eigenvalue of the matrix $\underline{\underline{\Lambda}}^{LS}$. But by Eq. (62) there exists a map from the eigenvalues of $\underline{\underline{\gamma}}$ to those of $\underline{\underline{\Lambda}}^{LS}$ (26). It follows that the orbital energies are functions of the occupation numbers (or vice versa); thus the energy is in this approximation a functional of the one-matrix.

(65) $$E^{LS}(n_1 \cdots n_M; \varepsilon_1 \cdots \varepsilon_M; v) \rightarrow E^{LS}(n_1 \cdots n_M; v).$$

Finally, the commutation rules in Eq. (63) show that the natural spinorbitals obey an eigenvalue equation

(66) $$F^{LS} \chi_p = \mu_p \chi_p \; .$$

Comparison of Eq. (66) and (26) makes explicit the similarities between the exact one-matrix theory and the present, approximate one.

IV. D Eigenvalues of $\underline{\underline{F}}^{LS}$

In the exact MC-SCF theory based on Löwdin's equations it has been shown that the Lagrange multipliers satisfy the identity of Eq. (47) . Representing ion states obtained by removal of a single electron from the system in the form of Eq. (50) it follows that ionization potentials are obtained by solution of Eq. (53) . The identity of the eigenvalues of $\underline{\underline{F}}^{LS}$ should be now apparent. For comparison of Eqs. (41) and (60) shows explicitly that the matrix of multipliers $\underline{\underline{\Lambda}}^{L}$ goes over to the matrix $\underline{\underline{\Lambda}}^{LS}$ when the non-hermitian potential F_{op} is replaced by the hermitian F_{av} . We claim that if the approximation of V_{op} by V_{av} is to be meaningful then the eigenvalues

$$\det[\underline{\underline{F}}^{LS} - \nu \underline{\underline{I}}] = \det[\underline{\underline{\Lambda}}^{LS} - \nu \underline{\underline{\gamma}}] = 0 \tag{67}$$

must be approximations to ionization potentials of the system of interest. Interestingly enough, $\underline{\underline{F}}^{LS}$ is already in diagonal form in the natural spinorbital representation for the one-matrix. Thus, in constrast to the exact MC-SCF theory in which ionization potentials must be extracted by a matrix diagonalization, no such diagonalization need be performed in the approximation due to Löwdin and Slater.

IV. C Summary

On the basis of the material presented one concludes that there exists an approximate one-matrix energy functional which leads to the commutation rules satisfied by the exact theory. In both cases the effective potential is hermitian for the ground state, and the same operator for all orbitals. Both functionals yield eigenvalues equations satisfied by the natural spinorbitals. Thus the orbital energies, and the total energy itself are functions of the one-matrix.

In the exact one-matrix theory of Section II, the eigenvalues of $F[\gamma]$ are derivatives of the total energy with respect to the occupation numbers. But we have shown by explicit computation that the eigenvalues of the approximate F^{LS} only approximate these derivatives, since they explicitly correspond to energy differences between N and (N-1) -electron states. Thus the degeneracy of the eigenvalues of does not hold in the approximation considered here. That is, since the eigenvalues of the latter correspond to different processes by which a single electron can be removed from the system, they are to be expected to be generally different.

The minimum energy required to remove an electron from the parent in a ground-state to ground-state transition between N and (N-1) -electron systems may be regarded as the chemical potential in the approximate theory. Its value, which can be got (as in the exact MC-SCF theory) from the long-range behaviour of the natural spinorbitals (23) can be taken as arbitrarily close to the first ionization potential of the system of interest. In a basis of natural spinorbitals which have been renormalized to the occupation numbers

(68) $$\varphi_p \equiv \sqrt{n_p}\,\chi_p$$

the energy in the Löwdin-Slater approximation can be written in the familiar form shown below

(69) $$E^{LS} = \frac{1}{2} \sum_{\mu_p < \mu} (h'_{pp} + \mu_p) .$$

Here h'_{pp} is a matrix element of the one-electron operator in this renormalized basis, and μ_p is $\varepsilon_p^{LS} n_p^{-1}$.

V. SUMMARY

The existence of the Hohenberg-Kohn theorem implies the existence of energy functionals based on the reduced density matrix of order $1 \leq P \leq N$. The special case $P = N$, reproduces the original Schrödinger functional. The special case $P = 2$ leads to an energy functional based on a linear reduced hamiltonian, as is well-known. But the difficulty in application of requisite N-representability conditions has so far prevented efficient and complete solution of the variational principle associated with the two-matrix functional. The special case $P = 1$ was considered in this discussion; we have shown that the structure of the resulting variational equations is of some interest, particularly when the one-matrix equations are compared to the MC-SCF equations. Notwithstanding their similarities, these two methods are fundamentally different. The one-matrix theory is based on a hermitian self-consistent potential, whereas the MC-SCF theory is based on a non-hermitian potential. Thus the occupation numbers and orbital energies are simultaneously well-defined in the one-matrix theory but not in MC-SCF theory.

The commutation rules discovered for the one-matrix theory are entirely consistent with a view of an atom or molecule as represented by an ensemble of single-particle states whose energy and occupation are simultaneously well-defined. In fact, we view these commutation rules as fundamental to the entire theory of a one-matrix energy functional, a view supported by the early work by Husimi. In this regard, it makes sense to view the one-matrix theory as a starting point for a discussion of an internal "thermodynamics" of atoms and molecules.

The exact form of the one-matrix functional is presently unknown except in the limiting case represented by Hartree-Fock theory. One approximate functional was introduced by Slater and by Löwdin, and we have analysed it here. The functional is interesting in that it satisfies the commutation rules laid down for the exact functional, and can be manipulated to provide approximate ionization potentials of the system of interest. Investigation of the numberical results obtained by use of this approximate energy functional is under active investigation in this laboratory.

REFERENCES

1. P. Hohenberg and W. Kohn, Phys. Rev. 136, B864(1964).

2. An N-representable density is one which could be exhibited by a fermion system. A V-representable density is one which is in one-to-one correspondence with some local potential v.

3. T.L. Gilbert, Phys. Rev. 12, B2111 (1975).

4. M. Berrondo and O. Goscinski, Int. J. Quantum Chem. S9, 67 (1975).

5. R.A. Donnelly and R.G. Parr, J. Chem. Phys. 69, 4431 (1078).

6. R.A. Donnelly, J. Chem. Phys. 71, 2874 (1979).

7. M. Levy, Proc. Natl. Acad. Sci. USA 76, 6062 (1979).

8. S. Valone, private communication of a manuscript for publication in Phys. Rev. A.

9. S. Valone, J. Chem. Phys. 73, 1344 (1980).

10. J.E. Harriman, Phys. Rev. 17, A1249 (1978), ibid., A1257.

11. J. Katriel, C.J. Appellof, and E.R. Davidson, Int. J. Quantum Chem. 19, 293 (1981).

12. H. Primas and M. Sleicher, Int. J. Quantum Chem. 9, 855 (1975).

13. E. Kryachko, Int. J. Quantum Chem. 18, 1029 (1980).

14. R.G. Parr, R.A. Donnelly, M. Levy and W.E. Palke, J. Chem. Phys. 68, 3801 (1978).

15. R.T. Sanderson, Science 121, 207 (1055).

16. J.C. Slater, Phys. Rev. 81, 385 (1951).

17. A.J. Coleman, Rev. Mod. Phys. 35, 668 (1963), D.W. Smith, Phys. Rev. 147, 896 (1966).

18. P.-O. Löwdin, Phys. Rev. 97, 1474 (1955).

19. R. McWeeny, Rev. Mod. Phys. 32, 335 (1960), P.A.M. Dirac, Proc. Cambridge Phil. Soc. 26, 376 (1930).

20. K. Husimi, Physico-Mathematical Society of Japan, Proc. Ser. 3 22, 264 (1940).

21. J. Hinze and H.H. Jaffe, J. Am. Chem. Soc. 84, 540 (1962), J. Hinze, M.A. Whitehead, and H.H. Jaffe, ibid., 85, 148 (1963).

22. D.W. Smith and O.W. Day, J. Chem. Phys. 62, 113 (1975).

23. M.M. Morell, R.G. Parr, and M. Levy, J. Chem. Phys. 62, 549 (1975).

24. B.C. Carson and J.M. Keller, Phys. Rev. 121, 659 (1961).

25. R.A. Donnelly, manuscript submitted to Int. J. Quantum Chem.

26. The eigenvalues of a commuting pair of matrices can be related by an interpolating polynomial. Thus the eigenvalues of one matrix are functions of the eigenvalues of the other. This mapping is one-to-one in the case that the eigenvalues of one of the commuting pair are non-degenerate.

SELF-INTERACTION CORRECTION

John P. Perdew
Department of Physics and Quantum Theory Group
Tulane University, New Orleans, Louisiana, U.S.A. 70118

CONTENTS:

I. Introduction

An electron can interact coulombically with other electrons but not with itself. Approximations for many-electron systems which violate this simple principle are candidates for self-interaction correction.

The earliest density functional approach, the Thomas-Fermi approximation, was purged of self-interaction by Fermi and Amaldi.[1] Later self-consistent-field theories constructed the electron density $n(\mathbf{r})$ and energy E from orbitals $\psi_{\alpha\sigma}(\mathbf{r})$ and occupation numbers $f_{\alpha\sigma}$, where $\sigma = \uparrow$ or $\downarrow$:

$$n(\mathbf{r}) = \sum_{\alpha\sigma} n_{\alpha\sigma}(\mathbf{r}) \tag{1}$$

$$n_{\alpha\sigma}(\mathbf{r}) = f_{\alpha\sigma} |\psi_{\alpha\sigma}(\mathbf{r})|^2 \tag{2}$$

$$\int d^3r \, |\psi_{\alpha\sigma}(\mathbf{r})|^2 = 1 \tag{3}$$

$$0 \leq f_{\alpha\sigma} \leq 1 \tag{4}$$

$$\sum_{\alpha\sigma} f_{\alpha\sigma} = N, \tag{5}$$

where N is the total number of electrons. In the Hartree-Fock approximation, the electron-electron repulsion energy is made up of direct electrostatic

$$\frac{1}{2}\int d^3r \int d^3r' \, \frac{n(\underset{\sim}{r})\, n(\underset{\sim}{r}')}{|\underset{\sim}{r}-\underset{\sim}{r}'|} \tag{6}$$

and exchange

$$E_x = -\frac{1}{2}\sum_{\sigma}\sum_{\alpha,\alpha'} f_{\alpha\sigma} f_{\alpha'\sigma} \int d^3r \int d^3r' \, \frac{\psi^*_{\alpha\sigma}(\underset{\sim}{r})\psi^*_{\alpha'\sigma}(\underset{\sim}{r}')\psi_{\alpha'\sigma}(\underset{\sim}{r})\psi_{\alpha\sigma}(\underset{\sim}{r}')}{|\underset{\sim}{r}-\underset{\sim}{r}'|} \tag{7}$$

terms. The $\alpha' = \alpha$ terms in (7) constitute a self-exchange energy

$$-\frac{1}{2}\sum_{\alpha\sigma}\int d^3r \int d^3r' \, \frac{n_{\alpha\sigma}(\underset{\sim}{r})\, n_{\alpha\sigma}(\underset{\sim}{r}')}{|\underset{\sim}{r}-\underset{\sim}{r}'|} \tag{8}$$

which dominates the total exchange energy for atoms[2] (Table I). Thus the Hartree approximation, which retains only these terms, involves a self-interaction correction to (6). On the other hand, the Hartree-Fock approximation is naturally self-interaction free and requires no afterthought correction.

There are two distinct reasons why the Hartree-Fock approximation may be undesirable: (1) The exchange integrals like (7) are difficult. They can be approximated by functionals of the electron density, as in the $X\alpha$[3] or local density approximations[4], at the cost of introducing a self-interaction error. There are several schemes for correcting this error.[5-9] (2) The Hartree-Fock approximation neglects dynamical correlation which may be as important as exchange for the low-density valence electrons which determine many of the interesting chemical and physical properties of atoms, molecules and solids. The theorems of modern density functional theory[4, 10, 11] sanction the hope of including these effects within a self-consistent field calculation. Some approximate density functionals for exchange <u>and</u> correlation are naturally self-interaction free but rather difficult to implement,[12] while others like the local spin density (LSD) approximation, [4, 11] or its generalization as a gradient expansion,[13, 14] are easily-implemented but contain a self-interaction error. Again there are several competing brands of self-interaction correction[15-20]

Here we will review the theory and applications of a particular method of self-interaction correction known as SIC.[17-20] As far as we know, it is the only method which satisfies all four of the following properties: (1) It can be applied to correlation as well as exchange; the two are treated on an equal footing. (2) It could be used in order to correct nonlocal functionals as well as local ones, and would yield no correction to the exact functional. (3) It is fully self-consistent: the one-electron potentials are derived variationally from the total energy. (4) It is exact for isolated electrons, and gives the exact energy per electron in systems where the density varies slowly over space. For electronic states which are either localized or localizable by unitary transformation, SIC yields a number of improvements over LSD in the density, the total energy and its separate exchange and correlation pieces, and especially the orbital energy eigenvalues which more closely approximate physical removal energies. However, there are some practical difficulties in applying SIC to extended orbitals, which will also be discussed.

II. Theory

A. A Property of the Exact Exchange-Correlation Functional

We will work within the spin-density version[11] of density functional theory. The total density is the sum of up- and down-spin densities

$$n(\underset{\sim}{r}) = n_\uparrow(\underset{\sim}{r}) + n_\downarrow(\underset{\sim}{r}) \,, \tag{9}$$

and the energy functional is

$$E_v[n_\uparrow, n_\downarrow] = \int d^3r\, v(\underset{\sim}{r})\, n(\underset{\sim}{r}) + T[n_\uparrow, n_\downarrow]$$

$$+ \frac{1}{2}\int d^3r \int d^3r' \frac{n(\underset{\sim}{r})\, n(\underset{\sim}{r}')}{|\underset{\sim}{r}-\underset{\sim}{r}'|} + E_{xc}[n_\uparrow, n_\downarrow], \tag{10}$$

where T and E_{xc} are the non-interacting kinetic and the exchange-correlation energies respectively. The ground-state energy and spin-densities are found by minimizing (10) subject to the constraint $\int d^3r\; n(\underset{\sim}{r}) = N$.

We intend to prove that, if $n_{\alpha\sigma}(\underset{\sim}{r})$ of (2) is a one-electron density ($f_{\alpha\sigma} = N = 1$), then

$$\frac{1}{2}\int d^3r \int d^3r' \frac{n_{\alpha\sigma}(\underset{\sim}{r})\, n_{\alpha\sigma}(\underset{\sim}{r}')}{|\underset{\sim}{r}-\underset{\sim}{r}'|} + E_{xc}[n_{\alpha\sigma}, 0] = 0, \tag{11}$$

i.e. the direct electrostatic self-interaction of an orbital is precisely cancelled by its exchange-correlation self-interaction. The proof is immediate if $n_{\alpha\sigma}(\underset{\sim}{r})$ is a possible ground-state density for some local external potential v(r), since in this case the left-hand side of (10) identically equals the sum of the first two terms on the right.

The exchange energy of (7) satisfies

$$\frac{1}{2}\int d^3r \int d^3r' \frac{n_{\alpha\sigma}(\underset{\sim}{r})\, n_{\alpha\sigma}(\underset{\sim}{r}')}{|\underset{\sim}{r}-\underset{\sim}{r}'|} + E_{x}[n_{\alpha\sigma}, 0] = 0, \tag{12}$$

so (11) amounts to a statement that the self-correlation of an orbital is zero:

$$E_c[n_{\alpha\sigma}, 0] = 0. \tag{13}$$

A general proof of the obvious results (11) and (13) will require an appeal to Levy's "constrained search" approach[21] to density functional theory:

Define the universal functional

$$Q[n_\uparrow, n_\downarrow] = \min \langle \hat{T} + \hat{V}_{ee} \rangle , \tag{14}$$

which searches the set of all antisymmetric wavefunctions yielding the given spin densities and delivers the minimum expectation value for the sum of the kinetic and electron-electron repulsion operators. Clearly

$$E_v[n_\uparrow, n_\downarrow] = Q[n_\uparrow, n_\downarrow] + \int d^3r\, v(\underset{\sim}{r})\, n(\underset{\sim}{r}) . \tag{15}$$

Similarly define

$$T[n_\uparrow, n_\downarrow] = \min \langle \hat{T} \rangle \tag{16}$$

and

$$E_{xc}[n_\uparrow, n_\downarrow] = Q[n_\uparrow, n_\downarrow] - T[n_\uparrow, n_\downarrow] - \frac{1}{2}\int d^3r \int d^3r' \frac{n(\underset{\sim}{r})\, n(\underset{\sim}{r}')}{|\underset{\sim}{r} - \underset{\sim}{r}'|} . \tag{17}$$

For a one-electron system, $\hat{V}_{ee} = 0$ so Q = T and (11) follows from (17). The one-electron density $n_{\alpha\sigma}(\hat{r})$ does not have to be a possible ground-state density.

B. The SIC Method

Many simple approximations for the exchange-correlation energy fail to satisfy the exact result (11). Consider for example the local spin density approximation

$$E_{xc}^{LSD}[n_\uparrow, n_\downarrow] = \int d^3r\, n(\underset{\sim}{r})\, \varepsilon_{xc}(n_\uparrow(\underset{\sim}{r}), n_\downarrow(\underset{\sim}{r})) , \tag{18}$$

where $\varepsilon_{xc}(n_\uparrow, n_\downarrow)$ is the exchange-correlation energy per particle of an electron gas with uniform spin densities $n_\uparrow$ and $n_\downarrow$. This approximation becomes exact for densities that vary slowly over space, but in general a local functional cannot precisely cancel a nonlocal one to satisfy (11).

The most accurate density functional calculations are performed in the Kohn-Sham scheme,[4] in which the kinetic energy functional T is treated exactly by the introduction of auxiliary orbitals and occupation numbers:

$$T = \sum_{\alpha\sigma} f_{\alpha\sigma} \int d^3r\, \psi_{\alpha\sigma}^*(\underset{\sim}{r}) \left(-\tfrac{1}{2}\nabla^2\right) \psi_{\alpha\sigma}(\underset{\sim}{r}) . \tag{19}$$

The orbitals which minimize the total energy, including some approximation $E_{xc}^{approx}[n_\uparrow, n_\downarrow]$ for exchange and correlation, satisfy the self-consistent Schrödinger equation

$$\left\{-\tfrac{1}{2}\nabla^2 + v(\underset{\sim}{r}) + \int d^3r' \frac{n(\underset{\sim}{r}')}{|\underset{\sim}{r}-\underset{\sim}{r}'|} + v_{xc}^{\sigma,approx}(\underset{\sim}{r}; [n_\uparrow, n_\downarrow])\right\} \psi_{\alpha\sigma}(\underset{\sim}{r})$$

$$= \varepsilon_{\alpha\sigma} \psi_{\alpha\sigma}(\underset{\sim}{r}) , \tag{20}$$

where the exchange correlation potential is the functional derivative

$$v_{xc}^{\sigma,approx}(\underset{\sim}{r}; [n_\uparrow, n_\downarrow]) = \frac{\delta}{\delta n_\sigma(\underset{\sim}{r})} E_{xc}^{approx}[n_\uparrow, n_\downarrow] . \tag{21}$$

For example, in the LSD of Eq. (18)

$$v_{xc}^{\sigma,LSD}(\underset{\sim}{r}; [n_\uparrow, n_\downarrow]) = \frac{\partial}{\partial n_\sigma}\left[n\, \varepsilon_{xc}(n_\uparrow, n_\downarrow)\right] . \tag{22}$$

According to (11), each orbital should carry zero self-interaction. If $E_{xc}^{approx}[n_\uparrow, n_\downarrow]$ fails to satisfy this condition, define the self-interaction corrected exchange-correlation energy[18]

$$E_{xc}^{SIC} = E_{xc}^{approx}[n_\uparrow, n_\downarrow] - \sum_{\alpha\sigma}\left\{\frac{1}{2}\int d^3r\int d^3r' \frac{n_{\alpha\sigma}(\mathbf{r})\, n_{\alpha\sigma}(\mathbf{r}')}{|\mathbf{r}-\mathbf{r}'|} + E_{xc}^{approx}[n_{\alpha\sigma}, 0]\right\}, \tag{23}$$

which does satisfy (11) for each occupied orbital.

Now seek the orbitals which minimize

$$\tilde{E} = \int d^3r\, v(\mathbf{r})\, n(\mathbf{r}) + T + \frac{1}{2}\int d^3r\int d^3r' \frac{n(\mathbf{r})\, n(\mathbf{r}')}{|\mathbf{r}-\mathbf{r}'|} + E_{xc}^{SIC} \tag{24}$$

with T given by (19), subject to the constraint (3) of normalization. The Euler equation is

$$\frac{\delta}{\delta\psi_{\alpha\sigma}^*(\mathbf{r})}\left\{\tilde{E} - \sum_{\alpha'\sigma'} f_{\alpha'\sigma'}\, \varepsilon_{\alpha'\sigma'} \int d^3r' \,|\psi_{\alpha'\sigma'}(\mathbf{r}')|^2\right\} = 0 \tag{25}$$

or

$$\left\{-\frac{1}{2}\nabla^2 + v(\mathbf{r}) + \int d^3r' \frac{n(\mathbf{r}')}{|\mathbf{r}-\mathbf{r}'|} + v_{xc}^{\alpha\sigma\, SIC}(\mathbf{r})\right\}\psi_{\alpha\sigma}(\mathbf{r}) = \varepsilon_{\alpha\sigma}\,\psi_{\alpha\sigma}(\mathbf{r}), \tag{26}$$

where the exchange-correlation potential contains self-interaction corrections:

$$v_{xc}^{\alpha\sigma, SIC}(\mathbf{r}) = v_{xc}^{\sigma, approx}(\mathbf{r}; [n_\uparrow, n_\downarrow]) - \left\{\int d^3r' \frac{n_{\alpha\sigma}(\mathbf{r}')}{|\mathbf{r}-\mathbf{r}'|} + v_{xc}^{\uparrow, approx}(\mathbf{r}; [n_{\alpha\sigma}, 0])\right\}. \tag{27}$$

The orbital energy eigenvalues of (26) have a well-defined meaning. Differentiation of (24) with respect to the occupation number $f_{\alpha\sigma}$, holding the orbitals and the other occupation numbers fixed, gives

$$\frac{\partial \tilde{E}}{\partial f_{\alpha\sigma}} = \int d^3r\, \psi^*_{\alpha\sigma}(\underset{\sim}{r}) \left\{ -\frac{1}{2}\nabla^2 + v(\underset{\sim}{r}) + \int d^3r' \frac{n(\underset{\sim}{r}')}{|\underset{\sim}{r}-\underset{\sim}{r}'|} + v_{xc}^{\alpha\sigma,SIC}(\underset{\sim}{r}) \right\} \psi_{\alpha\sigma}(\underset{\sim}{r}) = \varepsilon_{\alpha\sigma} . \tag{28}$$

Starting from the orbitals that satisfy (3) and (26), a small variation of the $\{f_{\alpha\sigma}\}$ and the $\{\psi_{\alpha\sigma}\}$ (with the normalization of the latter preserved) results in a first variation of the energy

$$\delta\tilde{E} = \sum_{\alpha\sigma} \varepsilon_{\alpha\sigma}\, \delta f_{\alpha\sigma} . \tag{29}$$

Now seek the occupation numbers which minimize (24), subject to the constraints (4) and (5). Minimization of $\tilde{E}$ subject to the number-conserving constraint of (5) is equivalent to minimization without this constraint of $\tilde{E} - \mu N$ where μ is the chemical potential:

$$\sum_{\alpha\sigma} (\varepsilon_{\alpha\sigma} - \mu)\, \delta f_{\alpha\sigma} \geq 0 , \tag{30}$$

which is an inequality because of the remaining constraint (4). In light of (4), equation (20) implies an "aufbau principle" for the ground-state: All orbitals with $\varepsilon_{\alpha\sigma} < \mu$ have $f_{\alpha\sigma} = 1$, and those with $\varepsilon_{\alpha\sigma} > \mu$ have $f_{\alpha\sigma} = 0$. Fractional occupation ($0 < f_{\alpha\sigma} < 1$) is possible only for orbitals with $\varepsilon_{\alpha\sigma} = \mu$, if there are any.

C. Exact Properties of LSD Preserved by SIC

While LSD is not self-interaction free, it is a good approximation for many purposes because its self-interaction is fairly small: The direct electrostatic energy can be estimated from the close inequality[22]

$$\frac{1}{2}\int d^3r \int d^3r' \, \frac{n(\underset{\sim}{r})\, n(\underset{\sim}{r}')}{|\underset{\sim}{r}-\underset{\sim}{r}'|} \le 1.092\, N^{2/3} \int d^3r \; n^{4/3}(\underset{\sim}{r}) \, , \tag{31}$$

while the LSD exchange energy is[11]

$$E_x^{LSD}[n_\uparrow, n_\downarrow] = -0.9305 \int d^3r \left[n_\uparrow^{4/3}(\underset{\sim}{r}) + n_\downarrow^{4/3}(\underset{\sim}{r}) \right] . \tag{32}$$

For a single occupied orbital ($N = 1$, $n = n_\uparrow = n_{\alpha\sigma}$, $n_\downarrow = 0$), the near cancellation between (31) and (32) is evident. Correlation tends to assist this cancellation. For the more tightly-bound orbitals, correlation is much smaller than exchange and the residue of the cancellation between (31) and (32) is small but positive. Thus the self-interaction correction of (23), when applied to LSD, lowers the calculated total energy.

LSD is exact for slowly-varying densities. Consider the limit of jellium, an electron gas of uniform density. The solutions of the LSD single-particle equation (20) are plane waves, orbitals extended over a large volume Ω. From (31) and (32), the self-interaction corrections to the energy and one-electron potential from an orbital extended over volume Ω are of order $\Omega^{-1/3}$ and vanish as $\Omega\to\infty$. Thus plane waves are also self-consistent solutions of the SIC single-particle equation (26), and SIC gives the exact energy per electron in the limit of slowly-varying density (unless the SIC equations also admit other, localized solutions of lower energy).

LSD has been remarkably useful outside its domain of formal validity. It gives reasonable results for densities that do not vary slowly on the scale of the local Fermi wavelength or screening length, i.e. for nearly all densities of physical interest. This unexpected success has been attributed[23] to the fact that LSD satisfies an exact conservation law: the exchange-correlation hole around an electron represents a deficit of one electron, i.e. the electron and its hole together constitute a neutral object.

If the ground-state of the interacting system can be connected adiabatically to that of a noninteracting system with the same density, the exchange-correlation energy can be written as the electrostatic interaction of each electron with its positively-charged hole:[3, 24]

$$E_{xc} = \frac{1}{2}\int d^3r \int d^3r' \, \frac{n(\underset{\sim}{r})\, \rho(\underset{\sim}{r}, \underset{\sim}{r}')}{|\underset{\sim}{r}-\underset{\sim}{r}'|} , \tag{33}$$

where $\rho(\underset{\sim}{r}, \underset{\sim}{r}')$ is the density at $\underset{\sim}{r}'$ of the hole around an electron at $\underset{\sim}{r}$, and

$$\int d^3r' \, \rho(\underset{\sim}{r}, \underset{\sim}{r}') = -1 . \tag{34}$$

The LSD and SIC holes can be deduced from (18) and (23):

$$\rho^{LSD}(\underset{\sim}{r}, \underset{\sim}{r}') = \rho_u(n_\uparrow(\underset{\sim}{r}), n_\downarrow(\underset{\sim}{r}); |\underset{\sim}{r}'-\underset{\sim}{r}|) \tag{35}$$

$$\rho^{SIC}(\underset{\sim}{r}, \underset{\sim}{r}') = \rho_u(n_\uparrow(\underset{\sim}{r}), n_\downarrow(\underset{\sim}{r}); |\underset{\sim}{r}'-\underset{\sim}{r}|)$$

$$-\sum_{\alpha\sigma} \frac{n_{\alpha\sigma}(\underset{\sim}{r})}{n(\underset{\sim}{r})} \left[n_{\alpha\sigma}(\underset{\sim}{r}') + \rho_u(n_{\alpha\sigma}(\underset{\sim}{r}), 0; |\underset{\sim}{r}'-\underset{\sim}{r}|) \right] , \tag{36}$$

where $\rho_u(n_\uparrow, n_\downarrow; |\mathbf{r}' - \mathbf{r}|)$ is the hole in an electron gas with uniform spin densities $n_\uparrow$ and $n_\downarrow$. Since ρ_u satisfies the sum rule (34), so obviously does ρ^{LSD}. From (36) we find

$$\int d^3r' \, \rho^{SIC}(\mathbf{r},\mathbf{r}') = -\left[\sum_{\alpha\sigma} f_{\alpha\sigma} n_{\alpha\sigma}(\mathbf{r})\right] / n(\mathbf{r}) . \tag{37}$$

Under the condition of adiabatic connection, all the occupation numbers are zero or one and the right-hand side of (37) reduces to -1. Thus, like LSD, SIC satisfies the exact sum rule (34).

For tightly-bound orbitals, SIC gives a much more realistic description of the <u>shape</u> of the hole than LSD does.[20] Fig. 1 shows the exchange hole (no correlation) about an electron located at distance r from the nucleus in the neon atom, for two different values of r. The full curves are exact (from Eq. (7)), while the dashed and dotted curves represent the SIC and LSD approximations respectively. Note that the observation point $\mathbf{r}' = \mathbf{r} + \mathbf{R}$ lies on a line with the electron and the nucleus.

D. SIC Orbital Energies and Long-Range Behavior

Consider the physical removal energy, defined as the energy needed to remove an electron from orbital $\alpha\sigma$, allowing for relaxation of the other orbitals. According to (29), this is the integral of the orbital energy over the occupation number

$$\Delta\tilde{E}_{\alpha\sigma}\Big|_{rel} = -\int_0^1 df_{\alpha\sigma} \; \varepsilon_{\alpha\sigma}(f_{\alpha\sigma}) , \tag{38}$$

which might be approximated as $-\varepsilon_{\alpha\sigma}(\frac{1}{2})$ using a transition state of half occupancy.[3] Equation (38) is a general self-consistent field result, and applies to the LSD,[3, 25] Hartree and Hartree-Fock approximations as surely as to SIC, once a continuation of the total energy to the interval $0 \leq f_{\alpha\sigma} \leq 1$ has been chosen.

A "self-interaction free" theory is one which satisfies (11), with $n_{\alpha\sigma}(\underset{\sim}{r}) = f_{\alpha\sigma}|\psi_{\alpha\sigma}(\underset{\sim}{r})|^2$ and $0 \leq f_{\alpha\sigma} \leq 1$. This category includes not only SIC but also the Hartree and Hartree-Fock approximations in their most natural continuation (equations (6) and (7)) to fractional occupation numbers. In a self-interaction free theory, the orbital energy eigenvalue $\varepsilon_{\alpha\sigma}$ in (38) depends on $f_{\alpha\sigma}$ largely through physical relaxation effects, i.e. through the effect on orbital $\alpha\sigma$ of the relaxation of the other orbitals. This effect is easily described: When an electron is removed from an atomic orbital $\alpha\sigma$, i.e. when $f_{\alpha\sigma} \to 0$, the remaining orbitals are drawn closer to the nucleus, which becomes more effectively screened. This makes $\varepsilon_{\alpha\sigma}(\tfrac{1}{2}) > \varepsilon_{\alpha\sigma}(1)$, i.e. in a self-interaction free theory one expects that

$$\Delta E_{\alpha\sigma}\Big|_{rel} \lesssim -\varepsilon_{\alpha\sigma}(1). \tag{39}$$

Since the screening effect from physical relaxation is often small, the inequality (39) is often close to an equality. In a self-interaction free theory, a single ground-state calculation yields an estimate of all the physical removal energies.

In contrast the spurious self-interaction in the LSD one-electron potential gives $\varepsilon_{\alpha\sigma}$ a strong but spurious dependence on $f_{\alpha\sigma}$. Thus in LSD accurate removal energies can only be found by the more elaborate means of transition-state or change-in-selfconsistent-field calculations, In particular, LSD band-structure calculations are in error due to self-interaction.

It is not hard to show, using formulas like (31) and (32), that the LSD self-interaction error of the orbital energy eigenvalues is about six times the self-interaction error from that orbital in the total energy.[20] While the left-hand side of (31) tends to cancel the right-hand side of (32), their functional derivatives cannot cancel so effectively due to the different powers of n which appear (n^2 vs. $n^{4/3}$). Thus self-interaction corrections to the LSD one-electron potential (see equation (27)) can lower the orbital energy eigenvalue significantly.

Note that (38) is an equation for the <u>relaxed</u> energy difference. It does <u>not</u> imply that the unrelaxed energy difference is

$$\Delta E_{\alpha\sigma}\Big|_{unrel} = -\varepsilon_{\alpha\sigma}(1). \tag{40}$$

Koopmans' theorem (40) is true in the Hartree-Fock approximation, but in SIC there is a well-defined positive correction to the right-hand side of (40). At least for atomic orbitals, this positive correction tends to cancel the negative orbital relaxation energy, making the SIC eigenvalue at full occupancy a better approximation to the physical removal energy than to the unrelaxed one.[20] Contrary to long-standing opinion, the orbital energy eigenvalues of density functional theory <u>do</u> have physical significance--but only when the functional is self-interaction free.

We turn now to the long-range behavior of the one-electron potential and density in an N-electron atom of nuclear charge Z. Far from the nucleus, the one-electron potential tends to

$$-\frac{Z^*}{r}, \tag{41}$$

where $Z^* = Z - N + 1$ in a self-interaction free theory and $Z - N$ in LSD. The calculated electron density falls off as

$$[n(r)]^{\frac{1}{2}} = k r^{\beta} e^{-\alpha r} \tag{42}$$

where $\beta = Z^*/\alpha - 1$, $\alpha = (-2\epsilon_{max})^{\frac{1}{2}}$, and ϵ_{max} is the greatest energy eigenvalue of the occupied orbitals.[18] From (39),

$$I \lesssim -\mathcal{E}_{max} \tag{43}$$

where I is the first ionization potential. Thus

$$[n(r)]^{\frac{1}{2}} \leq k r^{\beta'} e^{-\alpha' r} \tag{44}$$

where $\beta' = Z^*/\alpha' - 1$, and $\alpha' = (2I)^{\frac{1}{2}}$. Eq. (44) with $Z^* = Z - N + 1$ agrees with rigorous bounds on the long-range behavior of the density.[26, 27] Self-interaction free theories are correct far from the nuclear, where LSD is not.

The worst self-interaction errors of LSD appear in the orbital energy eigenvalue, and not in the total energy or the electron density. Consider for example a one-electron system, in which the exact one-electron hamiltonian is given by the first two terms of (20). The remaining two terms

$$\int d^3r' \frac{n_{\alpha\sigma}(\underset{\sim}{r}')}{|\underset{\sim}{r}-\underset{\sim}{r}'|} + v_{xc}^{\sigma,\,approx}(\underset{\sim}{r};[n_{\alpha\sigma},0]) \tag{45}$$

constitute a spurious self-interacting potential, which is displayed for the LSD hydrogen atom in Fig. 2(a). Over most of the volume of the atom, this potential is large (about 6eV out of a total binding energy of 13.6eV) but approximately constant, so it contaminates the LSD orbital energy eigenvalue far more than it contaminates the density profile (Fig 2(b)). However at very large distances from the nucleus the spurious self-interacting potential falls slowly to zero (like 1/r), and so contaminates the long-range behavior of the density.

E. Is SIC a Density Functional Theory?

So far we have thought of SIC as a self-interaction correction to a density functional approximation for exchange and correlation, such as LSD. We could equally well think of SIC as the Hartree method corrected by an approximation for exchange and correlation between different orbitals. Like the Hartree method, SIC has an orbital-dependent potential (27) which can generate non-orthogonal orbitals. In the atomic calculations of section III, the orbital overlaps are always small and Schmidt orthogonalization has only negligible effect on the total energy, but there could be situations in which this non-orthogonality causes trouble.

Clearly SIC does not fit into the Kohn-Sham formalism of (20) in which the effective one-electron is independent of the orbital quantum numbers α. We will show[20] that, nevertheless, the self-consistent SIC method is a density functional approximation and is wholly sanctioned by the Hohenberg-Kohn theorem. (These conclusions and the arguments which support them apply with equal force to the Hartree and Hartree-Fock methods.)

As in section II.A, we appeal to the "constrained search" idea.[21] Note that the SIC total energy (24) is manifestly a functional of the orbitals and occupation numbers, and define the SIC total energy as a functional of the spin densities by

$$E^{SIC}[n_\uparrow, n_\downarrow] = \min \tilde{E}[\{f_{\alpha\sigma}\}, \{\psi_{\alpha\sigma}\}] , \qquad (46)$$

which searches all orbitals and occupation numbers yielding the given spin densities and delivers the minimum. Then for any choice of spin densities integrating to N electrons,

$$E^{SIC}[n_\uparrow, n_\downarrow] \geq \tilde{E}\ (\text{self-consistent}) \qquad (47)$$

where the right-hand side is (24) evaluated using the self-consistent orbitals and occupation numbers. For the self-consistent $n_\uparrow$ and $n_\downarrow$, (47) becomes an equality.

F. Problems and Promise of SIC in Extended Systems

The SIC method can be implemented easily in LSD calculations for atoms and perhaps small molecules. What problems may arise for more extended systems?

First consider the tightly-bound core orbitals in a crystal. These can be represented equivalently (under unitary transformation) by Bloch orbitals extended over the whole crystal or Wannier orbitals localized on individual atoms. The Hartree-Fock exchange energy (7) is invariant under unitary transformation of the occupied orbitals, but the SIC exchange-correlation energy (23) is not. The extended orbitals have no self-interaction and so can be self-consistent solutions of both the SIC equation (26) and the LSD equation (20). But the SIC equation (26) will also admit localized orbitals as self-consistent solutions, and the self-interaction correction will give these localized solutions a lower total energy and lower orbital eigenvalues.[18] These corrections, which are easily implemented, will improve the calculated total energy and band-structure.

The valence-electron problem can be separated from the core-electron problem via electron-ion pseudopotentials. *A priori* pseudopotentials for valence electrons can be based on either the LSD or SIC methods. The resulting bare pseudopotentials are usually very similar; the most significant differences between LSD and SIC pseudopotentials arise from differences in screening by the valence electrons.[28]

There are practical problems in the application of SIC to more extended orbitals, especially those in the partially-filled bands of metals. How in practice can (26) be solved without either the spherical symmetry of tightly-bound core orbitals of the Bloch symmetry usually attributed to valence orbitals? Development of an easily-implemented, self-interaction free method of band-structure calculation is one of the greatest problems now facing density functional theory. (For further discussion, see the Afterword.)

III. Applications

A. Preface

Self-consistent self-interaction corrections to LSD have been applied[19, 20, 28-30] with good success to a number of problems in atomic physics and to the band-structure of insulating crystals. These calculations have been performed in the central field approximation, in which all the orbital densities $n_{\alpha\sigma}(\underset{\sim}{r})$ are replaced by their spherical averages. The electron gas correlation energy input to (18) is a parametrization[20] which matches Ceperley's [31] accurate Monte Carlo numerical values for low and metallic densities onto the exact high density limit.[32] Thus discrepancies between theory and experiment should accurately reflect the errors of the LSD and SIC approximations.

Before describing the impressive improvements of SIC over LSD, it is worth noting that SIC is <u>not</u> a cure-all. <u>Some</u> of the errors of LSD have a true many-body origin which has no connection with self-interaction. Altered boundary conditions can produce split-off modes such as surface plasmons[14, 24] which are not going to be successfully described by the correlation energy of a uniform electron gas. There are also strong correlations between degenerate or nearly-degenerate orbitals that are not properly described by either LSD or SIC. Thus SIC does not correct the errors of the LSD description of s-d and s-p interconfigurational energies.[20, 33, 34] And for strongly positive ions, where all the orbitals in a shell are nearly degenerate with each other and energetically well-separated from the other shells, both LSD and SIC predict a qualitatively wrong behavior (ℓnZ) for the correlation energy as the nuclear charge $Z \to \infty$, instead of the correct behavior (Z^0 or Z^1 depending on N).[30]

We do not know of any problems for which SIC results are significantly worse than LSD. We review here some problems for which they are significantly better. We start with those objects of traditional veneration in density functional theory, the total energy and electron density. Next come the orbital energy eigenvalues and band-structure, which assume a new physical significance in SIC. Finally we consider negative ions and autodetaching states, which become accessible to realistic self-consistent calculation only within a self-interaction free, correlated theory.

B. Total Energies of Atoms

The "experimental" total energy of an N-electron atom is the sum of the ionization potentials

$$\sum_{i=1}^{N} I_i \, , \tag{48}$$

corrected for relativistic and reduced-mass effects. In Table II we compare "experimental" total energies[35] with Hartree-Fock,[36] LSD and SIC values.[20]

The Hartree-Fock values are of course higher than experiment, and the LSD values are higher still, with an error about twice that of Hartree-Fock. The SIC values, on the other hand, are lower than experiment, with an error magnitude far less than that of Hartree-Fock for the lightest atoms but comparable to that of Hartree-Fock for Ne and Ar.

Exchange energies are compared in Table III. Taking the Hartree-Fock values[2] as a standard, we see that LSD consistently underestimates the magnitude of the exchange energy by 10-15%. SIC greatly improves the exchange energy, but tends to overestimate its magnitude slightly. This accounts for most of the error in the SIC total energy.

Correlation energies are compared with "experimental" values[35] in Table IV. LSD overestimates the magnitude of the correlation energy by as much as 100-200%. The SIC correlation energies,

$$E_c^{SIC} = \int d^3r \, n(\underset{\sim}{r}) \, \varepsilon_c(n_\uparrow(\underset{\sim}{r}), n_\downarrow(\underset{\sim}{r})) - \sum_{\alpha\sigma} \int d^3r \, n_{\alpha\sigma}(\underset{\sim}{r}) \, \varepsilon_c(n_{\alpha\sigma}(\underset{\sim}{r}), 0) , \quad (49)$$

overestimate the magnitude by only about 20%. Contrary to widespread opinion, the uniform electron gas correlation is relevant to small systems like atoms, once it has been purged of self-interaction.

Fig. 3 shows the LSD and SIC correlation energies for neutral atoms as a function of atomic number Z. Note the change of slope that occurs in the first transition series at chromium, as electrons of anti-parallel spin start to be added to the d shell. Except for such fine details, the LSD and SIC correlation energies for all atoms and ions may be described by a simple analytic model.[30]

C. Cohesive Energy of Metals

The cohesive energy is the small (1 - 6eV) difference of two large numbers: the energy of the free atom and the energy per atom of the solid. Although LSD-calculated total energies of atoms and solids are in error due to self-interaction, their difference yields a reasonable estimate of the cohesive energy in most metals.[37] In potassium for example, where the LSD self-interaction error of the atomic energy is 73eV, the LSD cohesive energy is only 1eV, in good agreement with experiment. The explanation is of course that nearly all the self-interaction error is carried by the ionic core, which is essentially the same for the solid as for the atom, while the self-interaction error of the K 4s orbital is much smaller that the cohesive energy.

The 3d transition metals are a notable exception to this rule. LSD gives accurate cohesive energies at both ends of the 3d series, but overestimates them by as much as 70% near the middle.[37]

As we cross the transition series from calcium to copper, the 3d band is progressively filled, and the orbitals shrink in response to the increasing nuclear charge. As a result the LSD self-interaction of each 3d orbital grows progressively, from about 0.2eV to about 1.2eV. These orbitals are not inert core orbitals until we reach copper; in fact they hybridize strongly with the 4s orbitals. It seems likely then that the s-d valence shell carries a greater LSD self-interaction in the transition atom than it does in the solid, leading to the LSD overestimation of the cohesive energy.[19, 20]

D. Electron Density in Atoms

As is evident in Fig. 2(b), self-consistent LSD electron densities are too diffuse. Self-interaction corrections to the one-electron potential counter this error by drawing the density in closer to the nucleus.

The polarizability of an atom describes its density response to a weak, uniform external electric field. In comparison with LSD, SIC polarizabilities tend to be slightly closer to experiment for the rare gas atoms, but slightly further away for the group II A divalents.[38]

In section III B we saw how SIC reduces the LSD correlation energies of atoms by about a factor of two, bringing them into much better agreement with experiment. If this is not just a fluke, we should expect a similar improvement in the density response to the correlation potential, i.e. the difference between the electron density calculated self-consistently with and without correlation.

Fig. 4 shows the density response to the correlation potential for the helium atom in LSD and SIC calculations.[20] This figure also shows the difference between configuration interaction[39] (CI) and Hartree-Fock densities, which can be regarded as the exact response to the correlation potential. The SIC results are clearly more realistic than LSD, although both approximations miss the CI tendency to deplete the density in the intermediate region 0.5 a.u. $\lesssim r \lesssim$ 1.0 a.u.

E. Orbital Energy Eigenvalues and Removal Energies

As discussed in section II D, the orbital energies (multiplied by -1) should approximate the relaxed energy changes due to electron removal in SIC, but not in LSD. This conclusion is supported by Table V, which compares the eigenvalue of the least-bound occupied orbital to the measured first-ionization potential[40] in a series of atoms. On the average, the SIC eigenvalues agree with experiment even better than the Hartree-Fock eigenvalues do. They seem to satisfy the inequality (39) almost as an equality. In contrast, the LSD eigenvalues only capture about 60% of the ionization potential.

Qualitatively similar conclusions apply when the ionization process leaves the atom in an excited state. Table VI displays a detailed comparison between ground-state orbital eigenvalues and relaxed energy differences due to hole creation in Ar. First consider the relaxed energy differences $\Delta E_{\alpha\sigma}|_{rel}$ in comparison with measured removal energies:[41] The SIC values show the best overall agreement with experiment, while the Hartree-Fock[41] values are somewhat in error for the outer orbitals (due to neglect of correlation) and the LSD values are slightly in error for the inner orbitals (due to self-interaction).

Next compare the orbital eigenvalues to the measured removal energies in Table VI: The Hartree-Fock[41] eigenvalues overestimate the removal energies for inner orbitals (due to neglect of relaxation and correlation) while the LSD eigenvalues seriously underestimate the removal energies from all orbitals. In contrast, the SIC eigenvalues yield a good estimate of the removal energies. Finally, note that the self-interaction free theories (Hartree-Fock and SIC) satisfy the inequality (39), and SIC satisfies it more nearly as an equality.

F. Band Gaps in Insulators

Neither the Hartree-Fock nor the LSD approximations are really satisfactory for band-structure calculations. The former predicts band gaps for insulators that are too large,[42] often by as much as a factor of 2-5 relative to experiment, while the latter yields band gaps that are too small[43] by 20-40%.

The LSD band gaps are too small because the states at the bottom of the gap, being typically more localized than those at the top of the gap, carry a greater positive self-interaction error in their orbital energy eigenvalues. Exactly the same error occurs in LSD calculations for the free atom.

We show in Table VII, along with the LSD band-structure error $\Delta E_{gap} = E_{gap}^{LSD} - E_{gap}^{expt}$ for rare-gas solids, the difference Δ_{SIC} between the LSD and SIC eigenvalues of the outermost atomic orbital forming the valence-band maximum (e.g. 2p, 3p, 4p for Ne, Ar and Kr respectively). The correction Δ_{SIC} is expected to be accurate since the p-like valence bands in the solid are atomic-like, narrow, and well-separated from the other bands. Δ_{SIC} was calculated using the same local exchange-correlation functionals used in the corresponding band-structure calculations.[43] Table VII clearly shows that the self-interaction of the p orbital accounts for almost 100% of the LSD error in the insulating gap of the solid.[19, 20] This result is also confirmed by an actual SIC band-structure calculation for solid argon.[20, 44]

G. Negative Ions

The negative ions are a sensitive test of any many-electron theory, since a large fraction of their binding energy is due to correlation. In LSD the one-electron equations (20) do not admit self-consistent solutions for negative ions:[45] The eigenvalue of the extra electron iterates to a positive value because of the incorrect long-range behavior of the one-electron potential discussed in section II D. (If however the extra electron is artificially confined to a region near the atom, the LSD total energy is lower than for the neutral atom, and the difference is a rough estimate of the true binding energy.[46]) This kind of misbehavior cannot occur in a self-interaction free theory like Hartree-Fock or SIC: Because of the inequality (39), the eigenvalue of the extra electron will always be negative if binding is allowed by the total energy.

Table VIII shows Hartree-Fock and SIC binding energies of negative ions, calculated from total energy differences,[19, 20] in comparison with experiment.[47] The Hartree-Fock values[48] grossly underestimate the binding, and in some cases predict an instability for the experimentally stable negative ions, while the SIC values are remarkably close to experiment, giving us additional confidence in the SIC description of correlation.

H. Relativistic SIC: The Electron Affinity of Gold

Correlation effects are always important in the binding energy of negative ions, but relativistic effects turn out to be equally important for the heavier negative ions.

The LSD and SIC approximations require that the z-component of an electron's spin be a good orbital quantum number, so these approximations do not blend well with the full Dirac equation. Koelling and Harmon[49] have proposed a modified Dirac equation, with the spin-orbit interaction averaged out, for use with LSD. This equation has been used to perform relativistic SIC calculations for Au and Au^-.[50] Since the only valence electrons are s electrons, the neglect of spin-orbit interaction is well-justified.

The electron affinity of gold has been calculated as the difference between SIC total energies of Au and Au^-. In the non-relativistic limit ($c \to \infty$), the calculated affinity is 1.5eV, about half of which comes from correlation energy. The relativistically-calculated affinity is 2.5 eV, in good agreement with the experimental value of 2.31 eV.[47] Thus the extra stabilization of s orbitals due to the relativistic mass increase applies even to the most loosely-bound valence orbitals.

I. Autodetaching States

The LSD and SIC methods admit self-consistent solutions which look like excited electronic states as well as ground states. The excited-state solutions are not blessed by as many theorems as the ground-state ones, but they can still be quite realistic, as we saw for neutral atoms in Table VI of section III E. Among the excited states of atoms and ions are some which are energetically permitted to decay by emission of an electron--the autodetaching states. For example, the autodetaching states of a negative ion A^- show up as resonances in the scattering of electrons from the neutral atom or molecule A.

Table IX compares Hartree-Fock and SIC total energies for the lowest-lying $(ns)^2\ {}^1S$ resonances in H^-[29] with "exact" values calculated from close-coupling equations.[51] No self-consistent LSD solutions were found. The Hartree-Fock solutions place each $(ns)^2$ resonance of H^- above the threshold for excitation of the $(ns)^1$ state in H, while the SIC solutions place each resonance below threshold, in good agreement with the "exact" results. SIC also predicts the existence of other known resonances (e.g. $(2p)^2$, $(3p)^2$), and admits no self-consistent solution for those non-resonant configurations (e.g. $(1s)^1\ (2s)^1$) which were tried.

Autodetaching states can be regarded as ghostly remnants of bound states which have moved up into the continuum due to a weakening of the nuclear attraction. Consider for example an ion with $N \leq 2$ electrons and variable nuclear charge Z. Fig. 5 shows the total energy $\tilde{E}(Z, N)$ calculated by the SIC method. By (28), the slope of $\tilde{E}$ vs. N gives the eigenvalue of the orbital being populated ($1s\uparrow$ for $N \leq 1$, and $1s\downarrow$ for $1 < N \leq 2$). At $Z = 1$, the $(1s)^2$ configuration lies below the $(1s)^1$ configuration, with an energy difference which correctly approximates the binding energy of H^-. As Z is reduced, the binding energy of the second electron decreases until it vanishes at $Z = 0.93$, where the $(1s)^2$ configuration becomes an autodetaching state.[29]

Acknowledgements

Special thanks are due to Alex Zunger, who authored many of the ideas and most of the calculations presented here, and to my other collaborators Jim Rose, Herb Shore, Lee Cole, Mel Levy and Ed McMullen, who also made major contributions to this work. The support of the National Science Foundation and the gracious hospitality of the organizers and hosts of this conference at U.N.A.M. are also gratefully acknowledged.

Afterword: A Prediction of Future Trends, and a Comment on the Xα Method

The SIC and related methods have many easy and useful applications to atomic-like systems,[52] but are too cumbersome for widespread applications to more extended systems. For large molecules or condensed matter, SIC still points a finger toward the future: The exact density functional is strictly self-interaction free, and leads to Kohn-Sham one-electron equations in which the orbital energy eigenvalue may closely approximate the physical removal energy.

The only easily-implemented density functional approximations for exchange and correlation are local or semi-local ones, involving the local density or low-order derivatives of it. The density gradient expansion[13, 14] is such a semi-local approximation. It is valid for densities varying slowly over space, but the gradient terms violate the sum rule of Eq. (34).[14] A semi-local generalization[53] of the gradient expansion which corrects this error promises to yield highly accurate total energies in atoms, molecules and solids.

While local or semi-local approximations can yield accurate exchange-correlation energies E_{xc} which are approximately (not exactly) self-interaction free, the functional derivatives of these approximations lead to exchange-correlation potentials v_{xc}^{σ} which still contain serions self-interaction errors. There is a similar situation in the approximation of ordinary functions: The sine  can be reasonably approximated by square wells  , but the derivative of the sine  is not so reasonably approximated by the derivative of the square wells  ; it would be more consistent with the original choice to approximate the derivative by square wells  also. The self-consistency condition of (21) could be the Gordian knot of density functional theory: The trend of the future may be the separate semi-local approximation of E_{xc} and v_{xc}^{σ}, in order to make both nearly self-interaction free. In fact the direct approximation of v_{xc}^{σ}, is already under consideration.[39]

As a simple example, we consider the local approximation of Slater's Xα method:[3]

$$E_{xc} = -\alpha \frac{9}{8}\left(\frac{6}{\pi}\right)^{1/3} \int d^3r \left[n_{\uparrow}^{4/3}(\underset{\sim}{r}) + n_{\downarrow}^{4/3}(\underset{\sim}{r}) \right] \tag{A1}$$

$$v_{xc}^{\sigma}(\underset{\sim}{r};[n_{\uparrow},n_{\downarrow}]) = -\alpha \frac{3}{2}\left(\frac{6}{\pi}\right)^{1/3} n_{\sigma}^{1/3}(\underset{\sim}{r}) , \tag{A2}$$

where α is an adjustable parameter. Taking (31) as an equality, α can be chosen to eliminate the self-interaction either from the total energy E:

$$\alpha_E = \frac{8}{9}\left(\frac{\pi}{6}\right)^{1/3} 1.092 = 0.78 , \tag{A3}$$

or from the orbital energy eigenvalue ε:

$$\alpha_{\varepsilon} = \frac{3}{2}\alpha_E = 1.17 . \tag{A4}$$

The choice (A3) is similar to the value α = 2/3 derived by Kohn and Sham,[4] while the choice (A4) is more similar to the value α = 1 originally proposed by Slater[3] and often used in band-structure calculations on empirical grounds.

References

1. E. Fermi and E. Amaldi, Mem. Accad. Ital. 6, 117 (1934).

2. J. C. Slater and J. H. Wood, Int. J. Quantum Chem. 4, 3 (1971).

3. J. C. Slater, The Self-Consistent Field for Molecules and Solids. (McGraw-Hill, New York, 1974).

4. W. Kohn and L. J. Sham, Phys. Rev. 140, A1133 (1965).

5. R. D. Cowan, Phys. Rev. 163, 54 (1967).

6. D. A. Liberman, Phys. Rev. 171, 1 (1968).

7. I. Lindgren, Int. J. Quantum Chem. 5, 411 (1971).

8. B. H. Brandow, Adv. Phys. 26, 651 (1977).

9. M. Gopinathan, Phys. Rev. A 15, 2135 (1977); T. J. Tseng, S. H. Hong and M. A. Whitehead, J. Phys. B 13, 4101 (1980).

10. P. Hohenberg and W. Kohn, Phys. Rev. 136, B864 (1964).

11. U. von Barth and L. Hedin, J. Phys. C 5, 1629 (1972).

12. O. Gunnarsson, M. Jonson and B. I. Lundqvist, Phys. Rev. B 20, 3136 (1976); J. A. Alonso and L. A. Girifalco, Phys. Rev. B 17, 3735 (1978).

13. D. J. W. Geldart and M. Rasolt, Phys. Rev. B 13, 1477 (1976).

14. D. C. Langreth and J. P. Perdew, Phys. Rev. B 21, 5469 (1980).

15. G. W. Bryant and G. D. Mahan, Phys. Rev. B 17, 1744 (1978); G. D. Mahan, Phys. Rev. B 22, 3102 (1980) and Phys. Rev. A 22, 1780 (1980).

16. H. Stoll, C. M. E. Pavlidou and H. Preuss, Theoret. Chim. Acta 149, 143 (1978); H. Stoll, E. Golka and H. Preuss, ibid. 55, 29 (1980).

17. A. Zunger and M. L. Cohen, Phys. Rev. B 18, 5449 (1978).

18. J. P. Perdew, Chem. Phys. Lett. 64, 127 (1979).

19. A. Zunger, J. P. Perdew and G. L. Oliver, Solid State Commun. 34, 933 (1980).

20. J. P. Perdew and A. Zunger, Phys. Rev. B 23, 5048 (1981).

21. M. Levy, Proc. Natl. Acad. Sci. USA 76, 6062 (1979); invited talk at this conference.

22. R. S. Gadre, L. J. Bartolotti and N. C. Handy, J. Chem. Phys. 72, 1034 (1980).

23. O. Gunnarsson and B. I. Lundqvist, Phys. Rev. B 13, 4274 (1976).

24. D. C. Langreth and J. P. Perdew, Phys. Rev. B 15, 2884 (1977).

25. J. F. Janak, Phys. Rev. B 18, 7165 (1978).

26. M. M. Morrell, R. G. Parr and M. Levy, J. Chem. Phys. 62, 549 (1975).

27. Y. Tal, Phys. Rev. A 18, 1781 (1978); 21, 2186 (E) (1980).

28. A. Zunger, Phys. Rev. B 22, 649 (1980).

29. J. P. Perdew, J. H. Rose and H. B. Shore, J. Phys. B 14, L233 (1981).

30. J. P. Perdew, E. R. McMullen and A. Zunger, Phys. Rev. A 23, 2785 (1981).

31. D. M. Ceperley, Phys Rev. B 18, 3126 (1978); D. M. Ceperley and B. J. Alder, Phys. Rev. Lett. 45, 566 (1980).

32. M. Gell-Mann and K. A. Brueckner, Phys. Rev. 106, 364 (1957); S. Misawa, Phys. Rev. 140, A1645 (1965).

33. O. Gunnarsson and R. O. Jones, Solid State Commun. 37, 249 (1981).

34. L. A. Cole and B. N. Harmon (unpublished),

35. A. Veillard and E. Clementi, J. Chem. Phys. 49, 2415 (1968).

36. C. Froese Fischer, The Hartree-Fock Method for Atoms. (Wiley, New York, 1977).

37. V. L. Moruzzi, J. F. Janak and A. R. Williams, Calculated Electronic Properties of Metals. (Pergamon, New York, 1978).

38. E. Zaremba, invited talk at this conference.

39. S. Jagannathan, Ph.D. thesis, University of Georgia, Athens, Ga., 1979 (unpublished); D. W. Smith, S. Jagannathan and G. S. Handler, Int. J. Quantum Chem. S13, 103 (1979).

40. C. E. Moore, Nat. Bur. Stand. (U.S.) Ref. Data Ser. 34, 1 (1970).

41. P. S. Bagus, Phys. Rev. 139, A619 (1965).

42. A. B. Kunz, Phys. Rev. B 12, 5890 (1975).

43. See e.g. Refs. (27)-(31) of Ref. 20.

44. A. Zunger (unpublished).

45. K. Schwarz, Chem. Phys. Lett. 57, 605 (1978).

46. H. B. Shore, J. H. Rose and E. Zaremba, Phys. Rev. B 15, 2858 (1978).

47. H. Hotop and W. C. Lineberger, J. Phys. Chem. Ref. Data 4, 539 (1975).

48. E. Clementi and C. Roetti, A. Data Nucl. Data Tables 14, 177 (1974); L. C. Green, M. M. Mulder, M. N. Lewis and J. W. Woll, Phys. Rev. 93, 757 (1954).

49. D. D. Koelling and B. N. Harmon, J. Phys. C 10, 3107 (1977).

50. L. A. Cole and J. P. Perdew (unpublished).

51. G. Seiler, R. Oberoi and J. Callaway, Phys. Rev. A 3, 2006 (1971); J. Callaway, Phys. Rep. 45, 89 (1978).

52. In addition to the applications described above, there is a recent and interesting self-interaction corrected calculation of quantum defects in Rydberg atoms: J. A. Armstrong, S. S. Jha and K. C. Pandey, Phys. Rev. A 23, 2761 (1981). Self-interaction corrections have also been applied to bound states on impurity atoms in metals: L. A. Cole and W. E. Lawrence (unpublished).

53. D. C. Langreth and M. J. Mehl (unpublished).

Table I. Hartree-Fock total, exchange, and self-exchange energies for lighter inert gas atoms. The correlation energy is shown for comparison. (After Refs. 2 and 35.) Energies in eV.

Atom	Total	Exchange	Self-exchange	Correlation
He	-77.9	-27.9	-27.9	-1.1
Ne	-3,497.8	-329.5	-269.2	-10.4
Ar	-14,334.7	-821.3	-617.8	-19.9
Kr	-74,883.4	-2,561.9	-1,653.6	

Table II. Calculated ground-state total energies of atoms compared with experiment, in eV. (After Ref. 20). The Hartree-Fock (HF) calculations are spin-restricted.

Atom	HF	LSD	SIC	Expt.
H	-13.6	-13.0	-13.6	-13.6
He	-77.9	-77.1	-79.4	-79.0
Li	-202.2	-199.8	-204.2	-203.5
Be	-396.5	-393.0	-399.8	-399.1
B	-667.4	-662.5	-672.0	-670.8
N	-1,480.2	-1,472.7	-1,488.9	-1,485.3
F	-2,704.9	-2,696.6	-2,720.7	-2,713.5
Ne	-3,497.8	-3,488.9	-3,517.6	-3,508.1
Na	-4,404.2	-4,392.6	-4,426.1	-4,414.7
Mg	-5,431.5	-5,418.3	-5,456.4	-5,443.2
Al	-6,581.5	-6,566.0	-6,608.8	-6,594.0
P	-9,271.0	-9,251.1	-9,303.7	-9,285.1
Ar	-14,334.7	-14,310.5	-14,378.3	-14,354.6

Table III. Exchange energies of atoms, in eV. (After Ref. 20.)

Atom	LSD	SIC	HF
H	-6.9	-8.5	-8.5
He	-23.2	-27.9	-27.9
Ne	-297.6	-337.8	-329.5
Ar	-755.8	-842.4	-821.3
Kr	-2,407.5	-2,632.0	-2,561.9

Table IV. Correlation energies of atoms, in eV. (After Ref.20.)

Atom	LSD	SIC	Expt.
H	-0.6	-0.0	-0.0
He	-3.0	-1.5	-1.1
Be	-6.0	-3.1	-2.6
Ne	-19.9	-11.4	-10.4
Mg	-23.9	-13.6	-11.6
Ar	-38.4	-22.3	-19.9

Table V. Comparison of ε_{max}, the least negative eigenvalue of the of the occupied orbitals, with the measured first ionization potential I. Energies in eV. (After Ref. 20.)

Atom	$-\varepsilon_{max}$			I
	HF	LSD	SIC	Expt.
H	13.6	7.3	13.6	13.6
Li	5.3	3.2	5.4	5.4
Na	5.0	3.1	5.1	5.1
K	4.0	2.6	4.3	4.3
N	15.4	8.3	14.9	14.5
P	10.7	6.3	10.0	10.5
Cr	6.5	4.0	6.7	6.8
Mn	6.7	4.6	7.1	7.4
He	25.0	15.5	25.8	24.6
Ne	23.1	13.5	22.9	21.6
Ar	16.1	10.4	15.8	15.8
Kr	14.3	9.4	14.0	14.0

Table VI. Comparison of ground-state orbital eigenvalues $\varepsilon_{n\ell}$, relaxed energy differences $\Delta E_{n\ell}|_{rel}$, and measured electron removal energies in atomic argon. Energies in eV. (After Ref. 20.)

Hole state	$-\varepsilon_{n\ell}$			$\Delta E_{n\ell}\|_{rel}$			
$n\ell$	HF	LSD	SIC	HF	LSD	SIC	Expt.
1s	3,227.0	3,097.0	3,220.0	3,195.0	3,179.0	3,195.0	3,206.0
2s	335.0	294.0	316.0	325.0	310.0	311.0	
2p	260.0	230.0	257.0	249.0	248.0	251.0	249.0
3s	34.8	24.0	30.2	33.2	29.9	30.1	29.9
3p	16.1	10.4	15.8	14.8	15.7	15.8	15.8

Table VII. Band gaps in rare gas solids, in eV. The first two columns show the experimental and LSD values for the energy gap between the occupied and unoccupied single-particle bands. The third column shows ΔE_{gap}, the error of the LSD gap, and the last column shows Δ_{SIC}, the self-interaction error of the eigenvalue for the highest occupied state in the atom. (After Refs. 19 and 20.)

Solid	E_{gap} Expt.	E_{gap} LSD	ΔE_{gap}	Δ_{SIC}
Ne	21.4	11.2	-10.2	9.9
Ar	14.2	8.3	-5.9	5.8
Kr	11.6	6.8	-4.9	4.9
Xe	9.3			4.2

Table VIII. Binding energies of negative ions in eV (or electron affinities of the corresponding neutral atoms). (After Refs. 19 and 20.)

Ion	Hartree-Fock	SIC	Expt.
H^-	-0.3	0.7	0.75
O^-	-0.5	1.6	1.46
F^-	1.4	3.6	3.40
Cl^-	2.6	3.8	3.62

Table IX. Total energies in eV of the $(1s)^2\ ^1S$ ground-state and some $(ns)^2\ ^1S$ autodetaching states in H^-. The "exact" $(2s)^2$ and $(3s)^2$ energies are the results of close-coupling calculations. (After Ref. 29.)

Parent state (H)	Exact	Daughter state (H^-)	HF	SIC	Exact
1s	-13.61	$(1s)^2$	-13.28	-14.32	-14.36
2s	-3.40	$(2s)^2$	-3.29	-3.80	-4.03
3s	-1.51	$(3s)^2$	-1.45	-1.75	-1.88
4s	-0.85	$(4s)^2$	-0.81	-1.01	

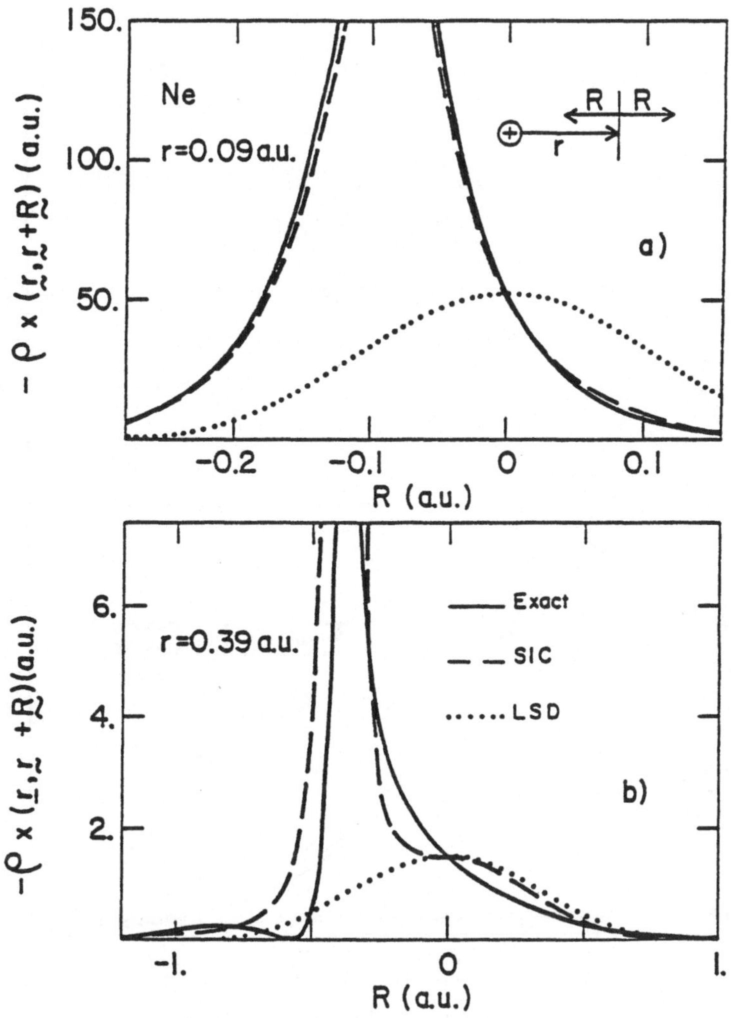

Fig. 1 Exchange hole in the neon atom. (From Ref. 20.)

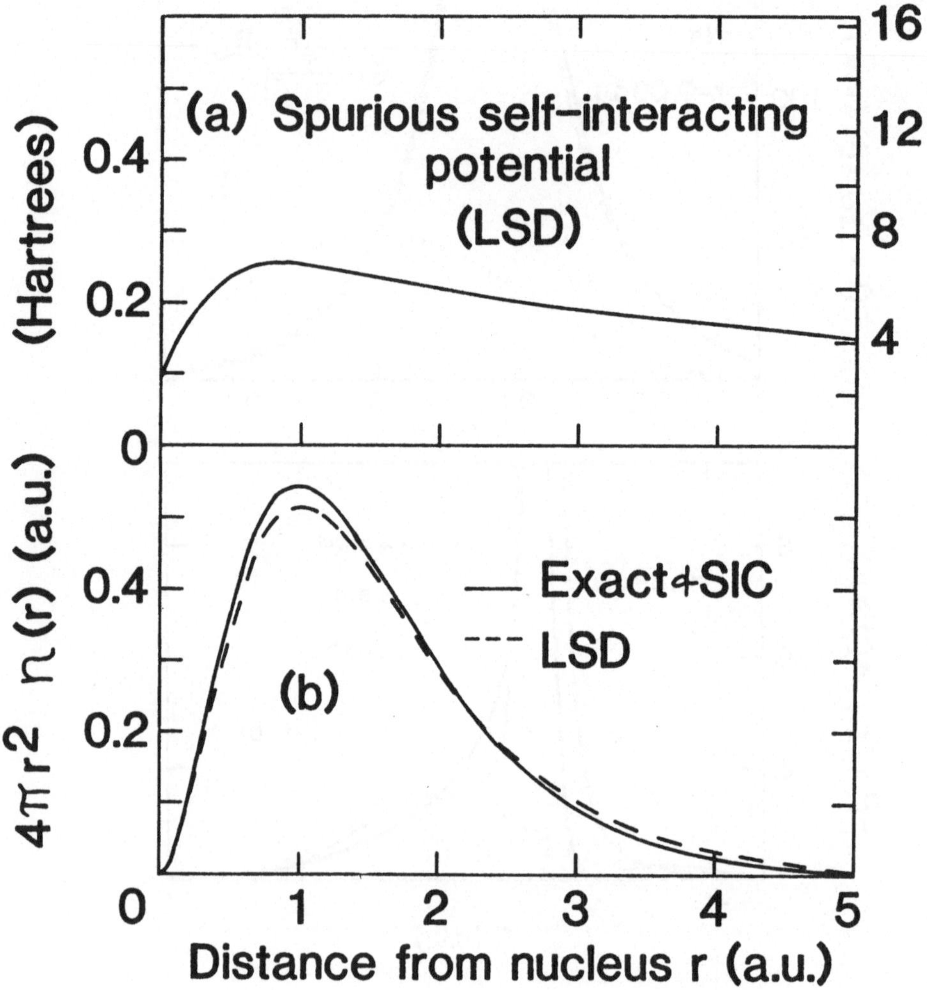

Fig. 2 The hydrogen atom. (a) Spurious self-interacting potential. (b) Radial distribution of the electron density.

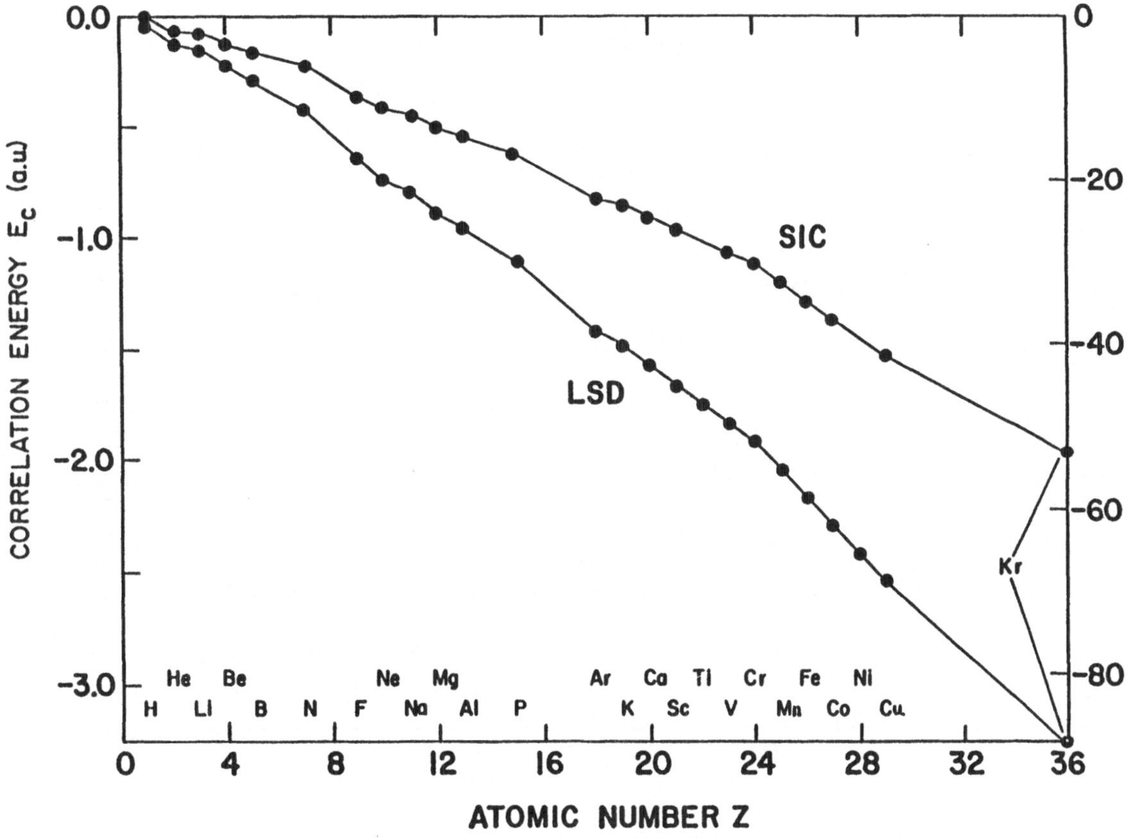

Fig. 3 Correlation energies of neutral atoms. (From Ref. 20.)

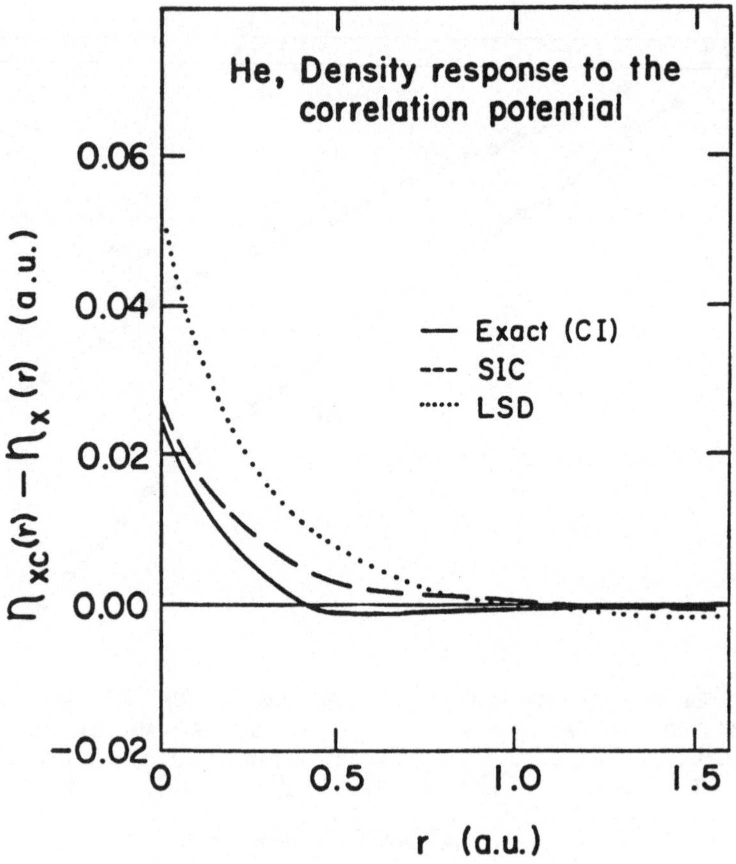

Fig. 4 Density response to the correlation potential in the helium atom. (From Ref. 20.)

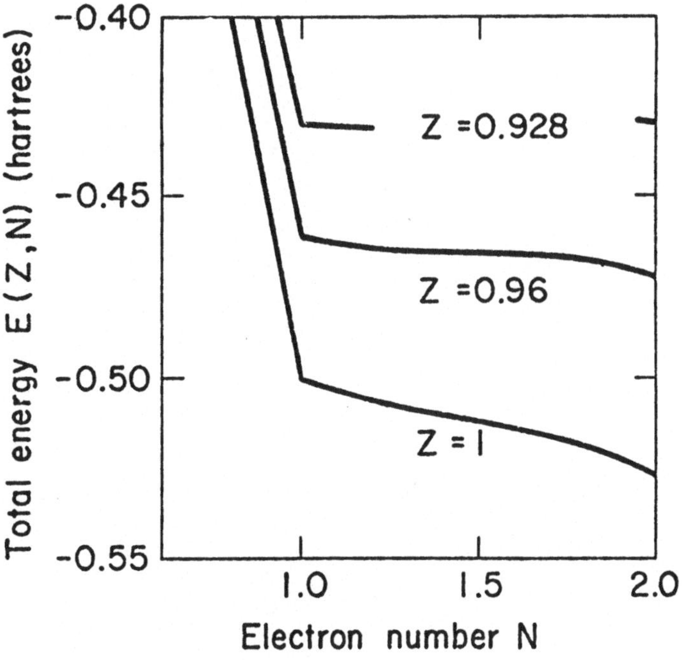

Fig. 5 Total SIC energy for $N \leq 2$ and $Z \leq 1$. (From Ref. 29.)

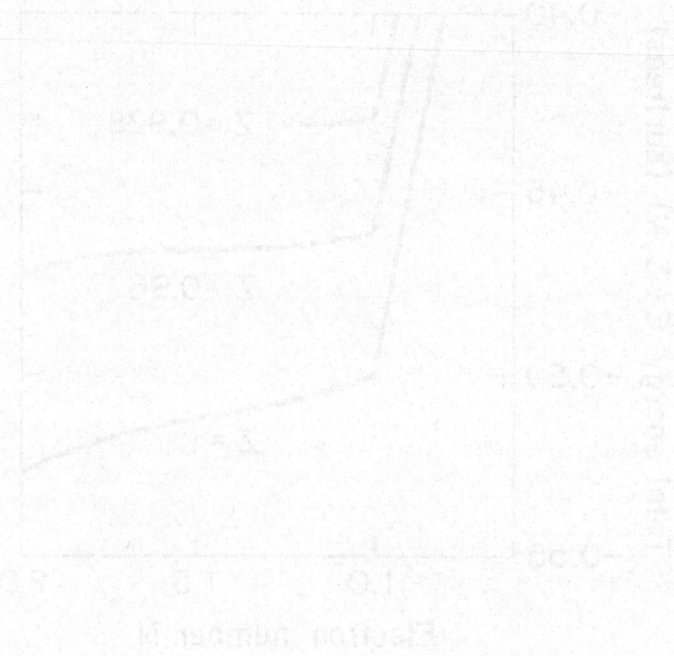

SOME RECENT DEVELOPMENTS IN DENSITY FUNCTIONAL THEORY

E. Zaremba
Department of Physics
Queen's University, Kingston, CANADA K7L 3N6

CONTENTS:

I. Introduction

Density functional theory [1,2] is now a popular (and cherished!) method for dealing with systems of many particles. Although primarily of importance in the theory of electronic structure, its philosophy has been adapted to problems as diverse as classical liquids[3] and nuclear structure.[4] It originated in the early statistical theory of atoms due to Thomas and Fermi[5] and, following a number of significant intermediate steps such as Slater's concept of statistical exchange[6], culminated in the fundamental theorem of Hohenberg and Kohn.[1] Since that time the basic theory has continued to evolve and has been applied extensively to a wide range of electronic problems. A number of excellent reviews have documented this progress.[7,8] However, in spite of its frequent successes, some fundamental questions still remain unanswered. These questions are the focus of much of the present work in the field and will be touched on briefly in the following. In addition there is also the pragmatic concern of facilitating the application of the theory to problems of current interest and, invariably, of increasing complexity. It is with this latter aspect that we shall be most concerned.

In the past, complex systems containing a large but finite number of particles have to a great extent been the reserve of quantum chemists who have been forced to develop effective means to deal with them. Solid state physicists have traditionally approached problems at the opposite extreme, being predisposed to infinite systems. Ironically, unpleasant, but essential, inconveniences such as boundaries, atomic disorder and a plethora of complex ordered compounds have tended to push solid state theory in the direction of quantum chemistry. But having been weaned on the theory of the uniform electron gas, it was natural for the solid state physicist to become an enthusiastic supporter of density functional theory when confronted with chemistry-related problems. Of course, this is not just a case of favouring a comfortable old shoe. The electron gas approach has provided new insights into old problems and its practical successes cannot be overstated.

It was with this perspective that we recently attempted to extend the applicability of the theory to situations which at first sight might appear unmanagably difficult.[9,10] Again the difficulties usually arise as a result of a complex geometry in which simplifying symmetries are not available. The electronic structure of defects in metals is a problem falling into this class and its study has motivated the concept of quasiatoms which appears to have many useful applications.[10] A quantitative definition of quasiatoms is given in Sec. III and their properties will be discussed extensively. Furthermore, the study of the response of the quasiatom to changes in its electronic environment has led to the general topic of linear response theory within the density functional formalism.[11] This is the second subject of these lectures and is outlined in Sec. IV. Past and present work on these topics will be described and hopefully will encourage others to pursue the subjects further; neither study has yet been completed nor (fortunately!) have all applications been exhausted.

II Density Functional Theory

A. Formalism

For completeness we shall outline some of the basic results of density functional theory.[1,2] We are interested in a system of N interacting electrons which find themselves is an external potential $v(\vec{r})$ due to compensating positive charges. The Hamiltonian of the system is

$$H = T + U + V \tag{1}$$

where T is the kinetic energy operator, U is the electron-electron interaction and V represents the interaction of the electrons with the external potential, given explicitly by

$$V = \int d\vec{r}\ \hat{n}(\vec{r})\ v(\vec{r}). \tag{2}$$

The Hohenberg-Kohn theorem[1] then states that there is a one-to-one correspondence between the ground state electronic density, $n(\vec{r})$, and the external potential $v(\vec{r})$. The proof of the theorem makes use of the quantum variational principle for the ground state energy of a nondegenerate system and begins by asserting the converse: namely, that a different potential $v'(\vec{r})$ generates the same density. If the Hamiltonian for the primed potential is $H' = T+U+V'$ and the corresponding ground state wavefunction is Ψ' we have

$$E' = \langle\Psi'|H'|\Psi'\rangle \ < \ \langle\Psi|H'|\Psi\rangle = \langle\Psi|H+V'-V|\Psi\rangle$$

i.e.,

$$E' < E + \int d\vec{r}\ n(\vec{r})[v'(\vec{r})-v(\vec{r})]. \tag{3}$$

The same argument leads to

$$E < E' + \int dr\ n(\vec{r})\ [v(\vec{r}) - v'(\vec{r})]. \tag{4}$$

Eqs. (3) and (4) imply the impossiblity

$$E+E' < E+E', \tag{5}$$

which demonstrates the incorrectness of our assumption. Thus there is a unique potential $v(\vec{r})$for each physical density $n(\vec{r})$, and $v(\vec{r})$ can therefore be considered as a density functional. Since the ground state wavefunction Ψ is uniquely determined by $v(\vec{r})$ it too is a functional of the density, as is any derivable quantity such as the ground state energy. The fact that the energy can be considered a functional of n is the fundamental result of the HK theorem.

In addition, HK define the density functional

$$E_v[n] = \int d\vec{r}\ n(\vec{r})\ v(\vec{r}) + F[n] \tag{6}$$

which they show satisfies a variational principle. Here $F[n] = \langle\Psi|T+U|\Psi\rangle$ is, as stated above, a universal functional of the density. We note that for the true density, n,

$$\begin{aligned} E_v[n] &= \langle\Psi|H|\Psi\rangle \\ &< \langle\Psi'|H|\Psi'\rangle \\ &= E_v[n'] \end{aligned} \tag{7}$$

where n' is the density corresponding to a different potential v'. Thus the energy functional $E_v[n]$ attains its minimum value for the true ground state density.

These results are formally exact but of course are somewhat empty if the functional F[n] is not known explicitly. It is at this point that the theory of a uniform electron gas provides a useful guide. It is now conventional to extract the classical Coulomb energy from F[n],

$$F[n] = \frac{1}{2}\int d\vec{r} \int d\vec{r}\,' \frac{n(\vec{r})n(\vec{r}\,')}{|\vec{r}-\vec{r}\,'|} + G[n] \tag{8}$$

leaving a universal functional G[n] representing kinetic, exchange and correlation effects. This separation is fine in principle but can lead to problems when unavoidable approximations to G[n] are made. For a uniform gas of density $\bar{n}$,

$$G[\bar{n}] = E_k + E_{xc} \tag{9}$$

where E_k is the kinetic energy of a degenerate system of <u>non-interacting</u> Fermions and E_{xc} is the usual definition of the exchange-correlation energy, both terms being functions of $\bar{n}$. Using this analogy, Kohn and Sham[2] suggested that in general one can write

$$G[n] = T_s[n] + E_{xc}[n] \tag{10}$$

where $T_s[n]$ is defined as the kinetic energy of a <u>non-interacting</u> system of Fermions having the <u>same</u> density as the fully interacting system. With this assumption, it follows that the density $n(\vec{r})$ can be constructed as

$$n(\vec{r}) = \sum_{i=1}^{N} |\psi_i(\vec{r})|^2 \tag{11}$$

where $\psi_i(\vec{r})$ are the orbitals entering a single Slater determinant. Whether or not this is always possible has not been demonstrated in detail but physically it appears to be a most reasonable assumption in view of the tremendous flexibility offered by the single-particle orbitals. It should of course be emphasized that the above assumption does not imply that the true ground state wavefunction Ψ can be constructed from a single Slater determinant.

Given this assumption, the HK variational principle

$$\frac{\delta E_v[n]}{\delta n(\vec{r})} = 0, \qquad \text{with} \int d\vec{r}\; \delta n(\vec{r}) = 0 \tag{12}$$

leads to the condition

$$\int d\vec{r}\;\; \delta n(\vec{r})\; [v(\vec{r}) + \phi(\vec{r}) + v_{xc}[\vec{r};n] + \frac{\delta T_s[n]}{\delta n(\vec{r})}\;] = 0 \tag{13}$$

where $\phi(\vec{r})$ is the electrostatic potential

$$\phi(\vec{r}) = \int d\vec{r}'\; \frac{n(\vec{r}')}{|\vec{r}-\vec{r}'|} \tag{14}$$

(Note: in density functional theory, electrons have a positive charge!) and

$$v_{xc}[\vec{r};n] = \frac{\delta E_{xc}[n]}{\delta n(\vec{r})} \;. \tag{15}$$

In a truly noninteracting system, one would have instead

$$\int dr\; \delta n(\vec{r}) \left[v(\vec{r}) + \frac{\delta T_s[n]}{\delta n(\vec{r})}\right] = 0 \tag{16}$$

Thus, the effective potential

$$v_{eff}[\vec{r};n] \equiv v(\vec{r}) + \phi(\vec{r}) + v_{xc}[\vec{r};n] \tag{17}$$

acts as an external potential for the noninteracting reference system. Since the ψ_i's for a <u>noninteracting</u> system are obtained from a single-particle Schrödinger equation, the orbitals of the reference system must satisfy

$$-\frac{1}{2}\nabla^2\psi_i(\vec{r}) + v_{eff}[\vec{r};n]\psi_i(\vec{r}) = \varepsilon_i\psi_i(\vec{r}) \tag{18}$$

The effective potential as defined in (17) is a functional of the density and so Eqs. (11), (17) and (18) must be solved self-consistently. This is the reduction of the many-body problem to a single-particle problem first carried out by Kohn and Sham.[2]

It should be noted that v_{eff} is a <u>local</u> potential as regards its position variable. Thus the present scheme is analogous to Hartree theory and since it includes exchange and correlation at the same time, it is far simpler than Hartree-Fock theory. As noted by KS, one could equally well start with a HF reference system by extracting the exact exchange energy from E_{xc}, leaving E_c, but at the expense of a non-local potential operator. Either alternative will in principle lead to an exact solution of the problem but the former is much easier to carry out in practice. The detailed relationship between the Hartree and HF formulations has not yet been fully explored and is an area which will undoubtedly lead to improvements in the present theory. We are here referring to problems endemic to the popular version of DFT based on the local density approximation.[2]

B. Local Density Approximation (LDA)

If the density of the system is slowly varying, say on the scale of the Fermi wavelength λ_F, then each small element of the electronic system can be thought of as uniform, suggesting the following approximation to $E_{xc}[n]$:

$$E_{xc}[n] \approx \int d\vec{r}\; n(\vec{r})\; \varepsilon_{xc}(n(\vec{r})). \tag{19}$$

Here $\varepsilon_{xc}(n)$ is the exchange-correlation energy per particle for a uniform system of density n. This defines the LDA in applications to an arbitrary system. Systematic corrections to the LDA in the form of a gradient expansion have been developed[12] but it appears that for systems which are highly inhomogeneous, such as atoms or metallic surfaces, the gradient corrections do not improve the results over the LDA.[13] This is now explained[14] by the way in which the LDA incorporates certain physical features correctly, thereby ensuring a reasonable estimate of the energy. For details we refer to Ref. 8.

Within the KS formulation, the LDA translates into the following form for the effective potential:

$$v_{eff}^{LDA}(\vec{r}) = v(\vec{r}) + \phi(\vec{r}) + \frac{d}{dn}\,[n\varepsilon_{xc}(n)]_{n=n(\vec{r})}. \tag{20}$$

One limitation of the LDA is clearly illustrated in applications to atomic systems. Since the exchange potential is proportioned to $n(\vec{r})^{1/3}$, it is exponentially localized. Thus $v_{eff}^{LDA}(\vec{r})$ behaves asymptoticly as $-(Z-N)/r$ in contrast to the $-(Z-N+1)/r$ decay of the HF potential. This deficiency has the effect of making the LDA eigenvalues considerably smaller in magnitude than their HF counterparts, although the ground state density and other properties of neutral atoms appear not to suffer excessively. However, one important consequence of this behaviour is the instability of negative ions in the LDA[15,16] (ie., valence eigenvalues are "positive"). For example, a two electron atom is only stable for a nuclear charge $Z \gtrsim 1.2$.[15] The source of this problem is the improper cancellation between the self-energy $\frac{1}{2}\int d\vec{r}\int d\vec{r}\,'\,|\psi_i(\vec{r})|^2|\psi_i(\vec{r}\,')|^2|\vec{r}-\vec{r}\,'|^{-1}$, treated explicitly in the electrostatic term, and the statistical approximation to the exchange term in E_{xc}. Recent corrections for this self-interaction have been proposed[17,18,19] but are not entirely satisfactory in that they are introduced in an *ad hoc* fashion. Perhaps a more satisfying approach to this problem might be the formulation of a local HF theory in which exchange effects are treated as accurately as possible within the framework of a local effective potential. This idea was first suggested by R.T. Sharp and G.K. Horton[20] and has recently been developed by J.D. Talman and co-workers,[21] but not with density functional theory in mind. To illustrate the implications of this idea, we show in Fig. 1 the optimal local exchange potential for Be compared with the KS local exchange potential for the same atomic density. The local approximation qualitatively reproduces the behaviour of the optimal potential but misses the r^{-1} tail and is too large in the core region. The differences between the two potentials actually lead to a very small difference in the atomic densities. It would be most desirable to develop an exchange energy functional corresponding to the optimized local HF potential. Such a functional would have the virtue of including self-interaction corrections without the inconvenience of different potentials for different orbitals.

We should also mention other attempts to account for self-interaction effects in a more rigorous way.[22] These theories are on a somewhat firmer foundation in that they develop a non-local approximation to the exchange-correlation energy functional by building in necessary physical constraints. Their chief drawback, however, is that they are difficult to implement and only a few tests have been carried out.

C. Reflections

We shall not describe in any detail the various applications of DFT which have been made to date nor the results obtained, since this is covered adequately elsewhere.[7,8] Rather we shall focus on some of the formal problems and practical difficulties which arise in general. The choice of $E_{xc}[n]$ has been dealt with in the previous section where it was noted that the LDA is frequently used in the KS formulation. In spite of the qualitative and, in many cases, quantitative successes of the LDA in atomic,[23] molecular[24] and solid state physics[25] one should be cautioned against becoming complacent in its application. A number of failures of the LDA have been uncovered and have motivated the generalizations described above. However these improvements generally lead to computationally more difficult problems and it remains to be seen whether the simplicity of the LDA can be retained as the sophistication of the theory increases. This is currently one aspect being actively investigated.

A second aspect concerns the practical difficulty of obtaining solutions to the single-particle Schrödinger equation for the geometry of interest and the subsequent iteration of the KS equations to self-consistency. The topics discussed in the following sections deal with this second aspect for certain problems of interest in atomic and solid state physics.

Explicit numerical solutions of the KS equations are usually only feasible in those situations with sufficient symmetry to simplify the problem. For closed-shell atoms, for example, the effective potential is spherically symmetric and allows one to effect a separation of variables. For perfectly ordered solids, one can take advantage of the translational symmetry of the lattice; the use of Bloch's theorem greatly simplifies the calculation of electronic bandstates in any of the well-known computational schemes. The bandstructure of transition metals has been studied in this way with (perhaps surprisingly) good results for various observable properties (cohesive energy, lattice constant, spin polarization, etc).[25]

However, whenever the system is perturbed in such a way as to decrease its original symmetry, the numerical problems are compounded considerably. Two, of many possible examples in which this situation arises can be sited. The application of an electric field to an atom leads to a distortion, a dipole moment, for which only axial symmetry remains. A direct calculation of the ground state properties in this distorted configuration would be difficult,[26] but fortunately one can appeal to perturbation theory in order to simplify the calculations since for weak fields, only distortions linear in the field are required. In HF theory the determination of the distorted orbitals remains a formidable task[27] while in density functional theory the calculations, as we shall see, are relatively straightforward as well as conceptually appealing.

A second example concerns the calculation of the energy of an atom placed on a metal surface. This chemisorption problem has been studied using a variety of approaches.[28-30] One difficulty is related to the extended nature of the substrate and the associated infinite number of degrees of freedom. This can be circumvented by a truncation of the system to a finite cluster which can be treated as a molecular system.[30] Provided the cluster is sufficiently large one would expect it to be representive of the metallic situation but a check on the effect of cluster size is necessary. The alternative approach would incorporate the extended nature of the metallic states from the beginning and one might expect a first step to be the determination of the states of a semi-infinite solid in the absence of the adsorbed atom. To a certain extent this is actually the approach taken.[28] In a jellium model the surface is translationally invariant so that the states are plane-wave-like with a well-defined momentum parallel to the surface. In a more realistic model the atomic structure of the surface would be included with periodicity along the surface providing a helping hand. However, with the addition of the impurity neither representation is valid since the metallic states are scattered by the impurity and even though one knows that the perturbation is localized to the vicinity of the impurity, it is not immediately apparent how this physical aspect should simplify the analysis. One basically must start again with a new calculation of the electronic states for this new situation. Clearly it would be desirable to build on the clean substrate calculation, which is easier, and include the adatom in a subsequent step. It is precisely this philosophy which has been adopted in the formulation of the idea of quasiatoms. The same idea was developed independently by Nørskov and Lang[31]. We now turn to a precise definition of what is meant by this term.

II Quasiatoms[10]

A: Definition

Some motivation for the idea of quasiatoms was given in the previous section. Here we define in a concrete way what is meant within a density functional setting.

The objective is determining the effect of adding an impurity to a host system which is characterized by a ground state electron density $n_o(\vec{r})$ and a compensating positive charge distribution $n_o^+(\vec{r})$. The main emphasis will be on obtaining an estimate of the change in energy caused by the impurity rather than on more directly calculable quantities such as wavefunctions and densities. The energy is a quantity difficult to obtain accurately since it often involves small differences of large numbers.

The energy of the host system is given by

$$E_o = G[n_o] - \int d\vec{r}_1 \int d\vec{r}_2 \frac{n_o(\vec{r}_1) n_o^+(\vec{r}_2)}{r_{12}} + \frac{1}{2} \int d\vec{r}_1 \int d\vec{r}_2 \frac{n_o(\vec{r}_1) n_o(\vec{r}_2)}{r_{12}} \tag{21}$$

where the second term is the interaction with the external potential provided by $n_o^+(\vec{r})$. We now consider the effect of adding Z electrons together with an additional positive charge distribution $\Delta n_z^+(\vec{r}) = Z\delta(\vec{r}-\vec{R})$, which for definiteness we take to be an impurity nucleus. The total ground state density for the new situation can be expressed as

$$n_Z(\vec{r}) = n_o(\vec{r}) + \Delta n_Z(\vec{r}) \tag{22}$$

where the impurity screening cloud $\Delta n_Z(\vec{r})$ will of course depend on the specific host being considered and, for metallic hosts, will tend to be localized in the vicinity of the impurity. The neutral entity consisting of the nucleus and its screening cloud will be termed a quasiatom. Through the influence of the host, the quasiatom need not be a spherically symmetric entity.

The ground state energy for this new situation is

$$E_Z = G[n_Z] - \int d\vec{r}_1 \int d\vec{r}_2 \frac{n_Z(\vec{r}_1)[n_o^+(\vec{r}_2)+\Delta n_Z^+(\vec{r}_2)]}{r_{12}} + \frac{1}{2}\int d\vec{r}_1 \int d\vec{r}_2 \frac{n_Z(\vec{r}_1)n_Z(\vec{r}_2)}{r_{12}}$$
$$+ \int d\vec{r}_1 \int d\vec{r}_2 \frac{n_o^+(\vec{r}_1)\Delta n_Z^+(\vec{r}_2)}{r_{12}} \tag{23}$$

and the change in energy associated with inserting the impurity is given by

$$\Delta E = \Big\{G[n_Z] - G[n_o] - \int d\vec{r}_1 \int d\vec{r}_2 \frac{\Delta n_Z(\vec{r}_1)\Delta n_Z^+(\vec{r}_2)}{r_{12}}$$
$$+ \frac{1}{2}\int d\vec{r}_1 \int d\vec{r}_2 \frac{\Delta n_Z(\vec{r}_1)\Delta n_Z(\vec{r}_2)}{r_{12}}\Big\} + \int d\vec{r}\,\phi_Z(\vec{r})[n_o(\vec{r})-n_o^+(\vec{r})] \tag{24}$$

where

$$\phi_Z(\vec{r}) = \int d\vec{r}\,' \frac{[\Delta n_Z(r') - \Delta n_Z^+(\vec{r}\,')]}{|\vec{r}-\vec{r}\,'|} \tag{25}$$

is the electrostatic potential due to the quasiatom.

The quantity in curly brackets will be recognized, by analogy with a free atom, as representing the self-energy ΔE_{self} of the quasiatom. It can be shown that the energy density associated with this term is localized so that such an identification is meaningful. The remaining term in (24), ΔE_{es}, takes explicit account of the electrostatic interaction between the quasiatom and the unperturbed host.

The quasiatom density, Δn_Z, of course depends on the nature of the host into which the impurity is inserted. For a given impurity located at the origin of coordinates, say, the quasiatom density will be a functional of the host density $n_o(\vec{r})$. If this functional dependence could be established, then Eq. (24) offers a conceptually transparent prescription for obtaining the quasiatom energy. Our objective is therefore to identify the features of the host environment which determine the screening cloud. If, due to the localized nature of the quasiatom, only a few local host properties are relevant, then the approach will certainly be be useful. As we shall see, various applications appear to confirm this expectation.

B. Uniform Density Approximation

One of the fundamental host parameters which determines the quasiatom density is the local host electronic density $n_o(\vec{R})$. In the limit that the host is a uniform electron gas - a jellium - this is the only relevant host parameter. In this limit the impurity-host electrostatic interaction vanishes and (24) becomes

$$\begin{aligned}\Delta E &= \overline{\Delta E}(\bar{n}) \\ &\equiv G[n_Z] - G[\bar{n}] - \int d\vec{r}_1 \int d\vec{r}_2 \, \frac{\Delta n_Z(\vec{r}_1)\Delta n_Z^{+}(\vec{r}_2)}{r_{12}} \\ &\quad + \frac{1}{2}\int d\vec{r}_1 \int d\vec{r}_2 \, \frac{\Delta n_Z(\vec{r}_1)\Delta n_Z(\vec{r}_2)}{r_{12}}\end{aligned} \tag{26}$$

where $\bar{n}$ denotes the electron density of the uniform electron gas. If the electron density in the host is slowly varying on the scale of the quasiatom then it is reasonable to take

$$\Delta E \simeq \overline{\Delta E}(n_o(\vec{R})) \tag{27}$$

as an estimate of its energy. This approximation which uses the jellium quasiatom energy evaluated for the density of the host at the position of the impurity nucleus will be referred to as the uniform density approximation (UDA). In a later section, the UDA will be shown to be the leading term in a systematic gradient expansion of the quasiatom energy.

C. Quasiatom Energy Curves

In applying the UDA, the uniform electron gas energy curve $\overline{\Delta E}(\bar{n})$ is required. This quantity has been evaluated by various workers using the KS equations.[32-35] For this spherically symmetric problem, the electronic wavefunctions are taken as

$$\psi_i(\vec{r}) = R_{k\ell}(r)\, Y_{\ell m}(\theta,\phi) \tag{28}$$

where $Y_{\ell m}(\theta,\phi)$ is a spherical harmonic and $R_{k\ell}(r)$ is a solution to the radial Schrödinger equation (for positive energies, $\varepsilon = \frac{1}{2} k^2$) having the asymptotic form

$$R_{k\ell}(r) \underset{r\to\infty}{\sim} \sin[kr-\frac{1}{2}\ell\pi + \delta_\ell(k)]/kr$$

with $\delta_\ell(k)$ the energy dependent phase shift. In the absence of the impurity, $R_{k\ell}(r)$ reduces to the spherical Bessel function $j_\ell(kr)$. Thus the screening charge density is obtained by summing over all occupied bound and continuum states, yielding[32]

$$\Delta n_Z(r) = 2 \sum_{b=1}^{B} |\psi_b(\vec{r})|^2 + \frac{1}{\pi^2} \sum_\ell (2\ell+1) \int_0^{k_F} dk\, k^2 [R_{k\ell}^2(r) - j_\ell^2(kr)]. \tag{29}$$

The first term is the bound state contribution, the factor of two explicitly accounting for the spin degeneracy. k_F is the Fermi wavevector related to the density by $k_F^3 = 3\pi^2\bar{n}$. Eq. (29) can also be generalized to the spin-polarized situation in which case the up- and down-spin densities are different.

Given the wavefunctions $\psi_i(\vec{r})$, the corresponding eigenvalues ε_i and the density, the energy can be partitioned as

$$\overline{\Delta E}(\bar{n}) = \Delta E_c + \Delta T_s + \Delta E_{xc} \tag{30}$$

Here,

(i) the Coulomb energy is

$$\Delta E_c = 4\pi \int_0^\infty dr r \Delta n_Z(r)\ [Q(r)-Z] \tag{31}$$

with

$$Q(r) = 4\pi \int_0^r dr' r'^2 \Delta n_Z(r') \tag{32}$$

the total screening charge within a distance r of the nucleus;

(ii) the kinetic energy is

$$\Delta T_s = 2 \sum_{b=1}^{B} \varepsilon_b + Z\varepsilon_F - \frac{2}{\pi} \sum_\ell (2\ell+1) \int_0^{\varepsilon_F} d\varepsilon\, \delta_\ell(\varepsilon)$$

$$-4\pi \int_0^\infty dr r^2 n_Z(r)\ \Delta v_{eff}[r;n] \tag{33}$$

where the impurity charge is related to the Fermi energy phase shifts by the Friedel sum rule

$$Z = \frac{2}{\pi} \sum_{\ell} (2\ell+1)\ \delta_{\ell}(\varepsilon_F) \tag{34}$$

and

$$\Delta v_{eff}[r;n] = v_{eff}[r;n] - v_{eff}[\infty;n] \tag{35}$$

and (iii), the exchange-correlation energy is given within the LDA by

$$\Delta E_{xc} = 4\pi \int_0^{\infty} drr^2 [n_Z(r)\varepsilon_{xc}(n_Z(r)) - \bar{n}\varepsilon_{xc}(\bar{n})]. \tag{36}$$

In the calculations to be described, ε_{xc} is taken from the work of Gunnarsson and Lundqvist.[14]

The main difficulty in the point-charge-in-jellium problem is achieving self-consistency and arises because of the unstable inward and outward flow of charge which is driven by the long-ranged Coulomb force. A trick for stabilizing the iterative procedure was introduced by Manninenen et al[36] which ensures convergence in most cases. This device was used in our calculations.

The energy curves for H, He and the first row atoms are shown in Fig. 2 and 3. A well-defined correlation with chemical properties is evident. It should be noted that for the two inert atoms, He and Ne, the energy curve is monotonic and close to being linear in the metallic density range. The increasing energy with density reflects the closed-shell nature of the atom which tends to repel electrons in the gas from its vicinity.

A second qualitative observation concerns the energy minimum exhibited by Li, Be and, to a much larger degree, by B through F. The minimum is associated with the open-shell nature of these atoms which provides a positive affinity for additional electrons. The depth of the minimum is compared with the electron affinities in Table I. The deepest minimum occurs for F, consistant with its large electron affinity. The significance of the minimum within the UDA is that

The quasiatom seeks out a host density near the minimum to maximize its stability. A deeper understanding into the behaviour of the energy curve is provided by the following theorem.

D. Slope Theorem.

We now establish an important relation between the slope of the energy curve and the quasiatom electrostatic potential.[10] Our derivation closely follows that of Budd and Vannimenus.[37]

Consider a jellium sphere of radius R with a charge Z at its centre as shown in Fig. 4. The background density is

$$n_o^+(r;R) = \bar{n}\theta(R-r), \quad \bar{n} = N_+/(\tfrac{4\pi}{3} R^3). \tag{37}$$

The ground state energy of the system is

$$E_Z(R) = \langle\Psi_Z(R)|H(R)|\Psi_Z(R)\rangle \tag{38}$$

where we have explicitly displayed the functional dependences on the parameter R. The Hamiltonian of the system is

$$H(R) = H_o - \int d\vec{r}\; \hat{\phi}_Z(\vec{r}) n_o^+(r;R) \tag{39}$$

where H_o contains those terms (electron kinetic energy, etc.) which are independent of R and $\hat{\phi}_Z(\vec{r})$ is the electrostatic potential operator defined as in (25) but with the electron density operator $\hat{n}(\vec{r})$ replacing $\Delta n_Z(\vec{r})$.

According to the Hellman-Feynman theorem[38] the variation of $E_Z(R)$ with R is given by

$$\frac{\partial E_Z}{\partial R} = -\int d\vec{r}\; \langle\Psi_Z(R)|\hat{\phi}_Z(\vec{r})|\Psi_Z(R)\rangle \; \frac{\partial n_o^+(r;R)}{\partial R}$$

$$= -\int d\vec{r}\; \phi_Z(\vec{r}) \; [-\tfrac{3}{R}\,\bar{n}\theta(R-r) - \bar{n}\delta(R-r)]. \tag{40}$$

We therefore find that the density derivative of E_Z is given by

$$\frac{\partial E_Z}{\partial \bar{n}} = -\bar{\phi}_Z - \frac{4\pi R^3}{3}\,\phi_Z(R) \tag{41}$$

where $\bar{\phi}_Z$ is the average of $\phi_Z(r)$ over the jellium sample,

$$\bar{\phi}_Z = \int d\vec{r}\ \phi_Z(\vec{r}) \tag{42}$$

A similar result obtains in the absence of the charge Z and so we find

$$\frac{d\overline{\Delta E}}{d\bar{n}} = -\ \bar{\phi}_Z \tag{43}$$

neglecting unimportant surface terms. Here ϕ_Z is the quasiatom potential (25). As a trivial extension we see that the curvature of the energy curve is related to variations of the mean Hartree potential with electron gas density

$$\frac{d^2\overline{\Delta E}}{d\bar{n}^2} = -\ \frac{d\bar{\phi}_Z}{d\bar{n}}\ . \tag{44}$$

Such variations reflect the degree to which the quasiatom responds to changes in the ambient density.

It should be emphasized that the slope theorem (43) is an exact result independent of the specific approximations for the exchange-correlation energy functional used in the evaluation of $\Delta n_Z(r)$. As such it provides a useful internal check on the self-consistency of the impurity in jellium calculation. More importantly, a knowledge of the slopes at a given set of density points allows one to accurately interpolate $\overline{\Delta E}(\bar{n})$ between these points.

Since the Hartree potential $\phi_Z(r)$ exhibits Friedel oscillations which decay as $\cos(2k_F r+\phi)/r^3$, there is no difficulty in evaluating the mean potential as indicated in (42). Occasionally however it is preferable to view the mean potential interms of the screening charge density in which case we have

$$\bar{\phi}_Z = -\ \frac{(4\pi)^2}{6} \int_0^\infty dr r^4 \Delta n_Z(r). \tag{45}$$

This clearly indicates the sensitivity of $\bar{\phi}_Z$ to a redistribution of charge as sampled by the fourth moment. Because of similar Friedel oscillations in $\Delta n_Z(r)$ the integral in (45) is only conditionally convergent but a careful limiting procedure yields the same result as from (42).

Returning to a discussion of the energy curves presented in Fig. 3, we note that the position of the minimum occurs at a density for which the mean potential vanishes. Since the potential for a neutral atom is everywhere negative, its mean potential is also necessarily negative. However for the quasiatom, the density and potential are oscillatory and the sign of the mean potential is determined to a considerable degree by the position and magnitude of the first oscillation. A negative slope suggests an overscreening of the impurity which can be viewed as a precursor to the formation of a negative ion. This tendency is illustrated for H in Fig. 5 which shows a plot of Q(r) vs. r. For a free negative ion $\bar{\phi}_Z \to +\infty$ which accounts for the behaviour of the curves near $\bar{n} = 0$ in those cases for which a stable negative ion is known to exist. The observed minimum for nitrogen does not correlate with the electron affinity of the free atom N since the latter in some sense possesses a closed shell.

E. Corrections to the UDA

In most applications of interest the quasiatom resides in a host environment which is nonuniform. It then becomes important to be able to estimate the magnitude of the corrections to the UDA. These corrections are of two kinds: one is explicitly displayed in (24) and represents the electrostatic interaction of the quasiatom with the unperturbed host,

$$\Delta E_{es} = \int d\vec{r}\ \phi_Z(\vec{r})\,[n_o(\vec{r}) - n_o^{+}(\vec{r})] \qquad (46)$$

Its evaluation requires a knowledge of $\Delta n_Z(\vec{r})$ for the actual problem being studied. However as a first approximation which yields a non-vanishing correction one can use the $\phi_Z(r)$ as determined in a jellium calculation for the local host density $n_o(\vec{R})$. To the extent that non-spherical distortions of the quasiatom are small, this should provide a reasonable estimate.

The second correction is to the quasiatom self-energy itself and, in this case, is explicitly dependent on distortions of the screening cloud induced by the nonuniform host. If the host density $n_o(\vec{r})$ has a slow spatial variation in the vicinity of the quasiatom one might expect the corrections to take the form of a gradient expansion whose first few terms are

$$\Delta E_{self} = \overline{\Delta E}(n_o(\vec{R})) + a(n_o(\vec{R}))|\nabla n_o(\vec{R})|^2 + b(n_o(\vec{R}))\nabla^2 n_o(\vec{R}) \qquad (47)$$

The first term is just the expected UDA energy and the coefficients a(n) and b(n) account for a linear spatial variation and a local curvature of the host density, respectively. It should be noted that the situation for a gradient expansion is expected to be more favourable for the quasiatom than for the corrections to the LDA since there is an independent length scale set by the quasiatom itself which is relatively independent of the nature of the host.

Expressions for the coefficients a(n) and b(n) can be obtained by considering a situation in which the spatial variations of $n_o(\vec{r})$ are both slow and small. We therefore consider a host positive charge distribution $n_o^+(\vec{r}) = \bar{n} + \delta n_o^+(\vec{r})$ such that $|\delta n_o^+(\vec{r})/\bar{n}| \ll 1$. The electron density in this situation will be denoted by $n_o(\vec{r}) = n_o^{(o)} + n_o^{(1)}(\vec{r}) +..$ where the superscripts refer to the order in the perturbation $\delta n_o^+(r)$. The lowest order term $n_o^{(o)}$ is of course $\bar{n}$. The total energy of the host expanded to second order in δn_o^+ is

$$E_o = E_o^{(o)} - \int d\vec{r}_1 d\vec{r}_2 \, \frac{n_o^{(1)}(\vec{r}_1)\delta n_o^+(\vec{r}_2)}{r_{12}}$$

$$+ \frac{1}{2}\int d\vec{r}_1 \int d\vec{r}_2 \, \frac{\delta n_o^+(\vec{r}_1)\delta n_o^+(\vec{r}_2)}{r_{12}} \qquad (48)$$

Simarily, we consider the same perturbation applied to the system with the impurity charge Z present. The electron density $n_Z(\vec{r})$ has a similar expansion, $n_Z(\vec{r}) = n_Z^{(o)}(\vec{r}) + n_Z^{(1)}(\vec{r}) +...,$ as does the electrostatic potential $\phi_Z(\vec{r}) = \phi_Z^{(o)}(\vec{r}) + \phi_Z^{(1)}(\vec{r}) +....$ With these definitions the total energy of the host plus impurity system can be written to second order in the perturbation as

$$E_Z = E_o^{(o)} + \overline{\Delta E}(n_o^{(o)}) - \int d\vec{r}\, \phi_Z^{(o)}(\vec{r})\delta n_o^{+}(\vec{r})$$

$$- \frac{1}{2}\int d\vec{r}\, \phi_Z^{(1)}(\vec{r})\delta n_o^{+}(\vec{r}) + \frac{1}{2}\int d\vec{r}_1 \int d\vec{r}_2 \, \frac{\delta n_o^{+}(\vec{r}_1)\delta n_o^{+}(\vec{r}_2)}{r_{12}} . \quad (49)$$

The quasiatom energy correct to second order is obtained as the difference of (49) and (48),

$$\Delta E = \overline{\Delta E}\,(n_o^{(o)}) - \int d\vec{r}\, \phi_Z^{(o)}(\vec{r})\delta n_o^{+}(\vec{r}) - \frac{1}{2}\int d\vec{r}\, \phi_Z^{(1)}(\vec{r})\delta n_o^{+}(\vec{r}) \quad (50)$$

In passing, we note that one retrieves the slope theorem from Eq. (50) for a spatially uniform perturbation.

The reduction of Eq. (50) to the final desired form requires a number of steps. First we must isolate the electrostatic correction ΔE_{es} using the appropriate expansions for the potential and charge density,

$$\Delta E_{es} = \int d\vec{r}(\phi_Z^{(o)}(\vec{r}) + \phi_Z^{(1)}(\vec{r})+\ldots)(n_o^{(1)} + n_o^{(2)}(\vec{r})+\ldots-\delta n_o^{+}(\vec{r})) \quad (51)$$

The leading term here is the result anticipated in the discussion following Eq. (46). Using (51) we can rewrite Eq. (50) as

$$\Delta E = \Delta E_{es} + \overline{\Delta E}\,(n_o^{(o)}) - \int d\vec{r}\, \phi_Z^{(o)}(\vec{r})[n_o^{(1)}(\vec{r}) + n_o^{(2)}(\vec{r})]$$

$$- \frac{1}{2}\int d\vec{r}\, \phi_Z^{(1)}(\vec{r})n_o^{(1)}(\vec{r}) + \frac{1}{2}\int d\vec{r}\, \phi_Z^{(1)}(\vec{r})[\delta n_o^{+}(\vec{r})-n_o^{(1)}(\vec{r})]. \quad (52)$$

Next, we make use of the slope theorem to re-express $\overline{\Delta E}(n_o^{(o)})$ in terms of $\overline{\Delta E}\,(n_o(\vec{R}))$. This allows us to make contact with the original UDA and leads to the result

$$\Delta E = \Delta E_{es} + \overline{\Delta E}(n_o(\vec{R})) - \int d\vec{r}\, \phi_Z^{(o)}(\vec{r})[n_o^{(1)}(\vec{r}) + n_o^{(2)}(\vec{r})+\ldots-n_o^{(1)}(\vec{R})-n_o^{(2)}(R)]$$

$$- \frac{1}{2}\int d\vec{r}[\phi_Z^{(1)}(\vec{r})n_o^{(1)}(\vec{r}) - \frac{d\phi_Z^{(o)}(\vec{r})}{d\bar{n}} n_o^{(1)}(\vec{R})^2]$$

$$+ \frac{1}{2}\int d\vec{r}\, \phi_Z^{(1)}(\vec{r})[\delta n_o^{+}(\vec{r}) - n_o^{(1)}(\vec{r})]. \quad (53)$$

It is now clear that when the host density is uniform over the extent of the quasiatom, the final three terms in (53) do not contribute. It is these terms which lead to the gradient expansion. A further observation is that the inhomogeneities in the host density are sampled by the quasiatom potential as indicated by the third term. This has led to an alternative method for correcting the UDA.[10]

The final step is to employ a gradient expansion for the host density,

$$n_o^{(1)}(\vec{r}) = n_o^{(1)}(\vec{R}) + \sum_i r_i \nabla_i n_o^{(1)}(\vec{R}) + \frac{1}{2} \sum_{ij} r_i r_j \nabla_i \nabla_j n_o^{(1)}(\vec{R}) + \ldots \quad (54)$$

We must also deal with the quantity $\phi_Z^{(1)}(\vec{r})$ which is the first order change in the quasiatom potential. It is related to the perturbation by a linear response function according to

$$\phi_Z^{(1)}(\vec{r}) = \int d\vec{r}' K_Z(\vec{r},\vec{r}')\delta n_o^{+}(\vec{r}'). \quad (55)$$

Here the kernel is given by

$$K_Z(\vec{r},\vec{r}') = -\int d\vec{r}_1 \int dr_2 \frac{1}{|\vec{r}-\vec{r}_1|} [\chi_Z(\vec{r}_1,\vec{r}_2) - \chi_o(\vec{r}_1,\vec{r}_2)] \frac{1}{|\vec{r}_2-\vec{r}'|} \quad (56)$$

where χ_Z and χ_o are the density response functions for the system with and without the impurity, respectively. Using the expansion of the density in (53), together with (55), one can identify the coefficients of $|\nabla n_o^{(1)}(\vec{R})|^2$ and $\nabla^2 n_o^{(1)}(\vec{R})$ appearing in Eq. (47). One finds

$$a(n) = -\frac{1}{6} \int d\vec{r} \sum_\alpha r_\alpha \Phi_\alpha(\vec{r};n) \quad (57)$$

and

$$b(n) = -\frac{1}{6} \int d\vec{r}\, r^2 \phi_Z^{(o)}(\vec{r};n) \quad (58)$$

where

$$\Phi_\alpha(\vec{r};n) = \int d\vec{r}' r_\alpha' K_Z(\vec{r},\vec{r}') \quad (59)$$

We note that b(n) is simply defined as the second radial moment of the quasiatom potential as determined for a uniform electron gas according to the method outlined in Sec. IIIC. b(n) has been evaluated for He and has been used in some of the tests discussed in the following section.

The other coefficient, a(n) is more difficult to evaluate in that a calculation beyond the uniform gas problem is required. This is evident from (57) and (59) which indicate that a(n) is related to the dipole moment of the quasiatom induced by a linear variation in the background density. Specifically, we notice that

$$\delta\Delta n(\vec{r}) \equiv \delta n_Z(\vec{r})-\delta n_o(\vec{r}) = \int d\vec{r}_1 \int d\vec{r}_2 [\chi_Z(\vec{r},\vec{r}_1)-\chi_o(\vec{r},\vec{r}_1)] \frac{1}{|\vec{r}_1-\vec{r}_2|} \delta n_o^{+}(\vec{r}_2)$$
$$= \int d\vec{r}\,' [\chi_Z(\vec{r},\vec{r}\,')-\chi_o(\vec{r},\vec{r}\,')]\delta v_{ext}(\vec{r}\,') \qquad (60)$$

is the difference in the densities induced by an external potential between the impurity and pure host systems. Within the KS formulation, $\delta v_{ext}(\vec{r})$ generates a change in the effective potential, $\delta v_{eff}(\vec{r})$, appearing in Eq. (18) and determining the KS orbitals. However, since the KS system represents a collection of independent particles, the response $\delta n_Z(\vec{r})$ can equivalently be obtained using the independent-particle response function $\chi_Z^o(\vec{r},\vec{r}\,')$ with $\delta v_{eff}(\vec{r})$ as the total perturbation:

$$\delta n_Z(\vec{r}) = \int d\vec{r}\,' \chi_Z{}^o(\vec{r},\vec{r}\,')\delta v_{eff}(\vec{r}\,') \qquad (61)$$

The superscript 'o' is a reminder that the response function for independent particles is being considered. The change in the effective potential is given to lowest order in $\delta v_{ext}(\vec{r})$ by

$$\delta v_{eff}^{Z}(\vec{r}) = \delta v_{ext}(\vec{r}) + \int dr' \frac{\delta n_Z(\vec{r}\,')}{|\vec{r}-\vec{r}\,'|} + \int d\vec{r}\,' v'_{xc,Z}(\vec{r},\vec{r}\,')\delta n_Z(\vec{r}\,') \qquad (62)$$

with

$$v'_{xc,Z}(\vec{r},\vec{r}\,') = \left.\frac{\delta^2 E_{xc}[n]}{\delta n(\vec{r})\ n(\vec{r}\,')}\right|_{n=n_Z(\vec{r})} . \qquad (63)$$

The second term on the right-hand side in Eq. (62) is the change in the electrostatic potential due to the redistribution of electrons and the last term represents the change in the exchange-correlation potential. If the LDA is made, then v_{xc} is a function of the density and Eq. (63) becomes

$$v'_{xc,Z}(\vec{r},\vec{r}')_{LDA} = \frac{d}{dn} v_{xc}(n)\Big|_{n=n_Z(\vec{r})} \delta(\vec{r}-\vec{r}'). \tag{64}$$

Because $\delta v^{Z}_{eff}(\vec{r})$ is linearly dependent on $\delta n_Z(\vec{r})$, we see that Eq. (61) represents an integral equation for the induced density,

$$\delta n_Z(\vec{r}) = \int d\vec{r}'\chi_Z{}^{o}(\vec{r},\vec{r}')\delta v_{ext}(\vec{r}') + \int d\vec{r}' \int d\vec{r}''\chi_Z{}^{o}(\vec{r},\vec{r}')V_Z(\vec{r}',\vec{r}'')\delta n_Z(\vec{r}''), \tag{65}$$

where we have introduced $V_Z(\vec{r},\vec{r}') = \dfrac{1}{|\vec{r}-\vec{r}'|} + v'_{xc,Z}(\vec{r},\vec{r}')$.

A completely analogous result is obtained for the induced density in the absence of the impurity (Z=0),

$$\delta n_o(\vec{r}) = \int d\vec{r}'\chi_o{}^{o}(\vec{r},\vec{r}')\delta v_{ext}(\vec{r}') + \int d\vec{r}' \int d\vec{r}''\chi_o{}^{o}(\vec{r},\vec{r}')V_o(\vec{r}',\vec{r}'')\delta n_o(\vec{r}''). \tag{66}$$

The response function χ_o^o is in this case a function of $|\vec{r}-\vec{r}'|$ and so (66) can be solved for $\delta n_o(\vec{r})$ by a Fourier transformation. The difference between (65) and (66) generates an integral equation for the quantity $\delta\Delta n(\vec{r}) = \delta n_Z(\vec{r}) - \delta n_o(\vec{r})$ defined in (60). An explicit solution of the integral equation would allow one to evaluate a(n). Although work in this direction has begun no definitive results are available as yet. Fortunately there are some problems in which the gradient $|\nabla n_o|$ vanishes so that a(n) is not required.

F. Applications.

Most of the applications to date have made use of the UDA. A simple check on the UDA is provided by the calculation of the binding energy of an atom to a vacancy in jellium. The vacancy is defined simply by removing the positive background within a sphere of radius R such that $(4\pi/3)R^3\bar{n} = Z_v$, the valence of the

host being considered. The removal of this positive charge leads to a depletion of the electronic density near the vacancy.
This charge density is calculated self-consistently using the KS equations and represents the host into which the quasiatom is imbedded. An analogous calculation with the impurity charge at the centre of the vacancy can be performed in order to obtain the exact energy -within KS theory- of the impurity system. The binding energy, E_B, is then the difference between this energy and energy of the impurity when far from the vacancy, i.e., in a uniform electron gas.

The UDA estimate of the binding energy is given simply by

$$E_B = \overline{\Delta E}(n_o(\vec{R})) - \overline{\Delta E}(\bar{n}) \tag{67}$$

where $n_o(\vec{R})$ is the density at the centre of the vacancy in the absence of the impurity. This estimate is compared with the exact results in Table II for a number of impurity - host combinations. We observe that the trends obtained with the UDA agree with the correct behaviour and have a very simple interpretation in terms of the energy curves. The most straightforward case is that of He for which the energy curve increases monotonically, indicating the preference of the He atom for low density regions. Thus the observed binding of He to vacancies is consistent with its energy curve. Also shown in Table II are results for He which take into account the electrostatic and curvature corrections. The improvement over the UDA is evident and the final numerical values are typically within a few tenths of an electron volt of the correct value. This should not be taken as a general indicator of the reliability of the gradient expansion since He is probably a favourable case. However it is encouraging that the corrections do move the UDA estimate in the right direction.

Hydrogen on the other hand behaves in a qualitatively different way at low densities because of its tendency to form a negative ion. The minimum in the energy curve corresponds to a density at which the quasiatom is most stable and so the quasiatom is driven to those regions approaching the density at the minimum. If the gas density happens to be well to the right of the minimum, the lower density at the vacancy centre leads to binding. However, if the gas density is low corresponding to a point to the left of the minimum, a lowering of the density increases the quasiatom energy and binding to the vacancy will not occur. This explains why there is no binding of H to a Na vacancy, in agreement with the exact calculation, while the proton is bound by about 3eV to an Aℓ vacancy. However it should be emphasized that in a realistic model of the metal, the final conclusion concerning binding may be different. More sophisticated models (such as the spherical solid model)[39] in fact predict the opposite to be true in the case of Aℓ. Experimentally it is known that the positive muon, a light isotope of H ($m_H/m_\mu \approx 9$), is bound to vacancies by about 0.3eV. Evidently, a more detailed description of the electronic structure of the host beyond both the jellium and the spherical solid models is needed to account for this result.

The correct qualitative behaviour is also found for Li in jellium vacancies, but the agreement with the exact results is not as good as for H and He. It is not known yet whether gradient corrections can account for these discrepancies as they did in the case of He.

The chemisorption of hydrogen on metal surfaces can also be studied with the UDA and previous KS calculations for H on Aℓ and Na substrates provide an exact comparison.[29] The host in this case is the surface electron density profile which has been calculated by Lang and Kohn[41] for the jellium model. With this information the UDA gives the dependence of the quasiatom energy on its distance from the surface and the results are compared with the exact results in Fig. 6 . It is again remarkable how well the UDA accounts for the qualitative behaviour of the chemisorption energy for both high and low density substrates. Considering the complexity of the full KS calculations it is gratifying that such a simple approach works so well and offers some hope that the method can be exploited when considering even more complicated systems such as transition metals.

The UDA has also met with some success in accounting for the binding energy of molecular hydrides, MH, with M a group II metal atom. Here we have regarded the metal atom as the host, $n_o(\vec{r})$ being the free atom electron density. According to the UDA, all such molecules would be bound with an energy of $\sim$ 1.8 eV corresponding to the depth of the H energy curve at an equilibrium separation, R, determined by $n_o(R) = n_{min}$. The extent to which this prediction is verified can be seen by plotting the observed molecular binding energy and separation as a point in the energy-density plane (Fig. 2). The obvious clustering of the points in the vicinity of the energy minimum clearly suggests that the H quasi-atom is seeking that configuration which stabilizes its electron gas energy. Even for a highly inhomogeneous host like an atom, the local host density at the proton is a significant parameter contributing to the observed chemical properties of the combined system.

As a final example we briefly mention some preliminary work on interstitial potentials for H in simple metals, specifically Aℓ.[42] In order to apply the quasi-atom concept the Aℓ-metal host density must first be determined. It can be obtained from a self-consistent bandstructure calculation but a simpler method can be adopted if this is not available. The crudest approximation would be a superposition of free atom densities which however neglects the fact that the host ion is screened in a metallic environment and not by discrete valence states. This effect can be included in a refined approximation whereby one considers an isolated host ion at the centre of a vacancy in an otherwise homogeneous gas. The self-consistent calculation for the imbedded ion can be carried out as described previously (Sec. IIIC) and the resulting screening charge densities can be superposed at each lattice site. The final charge density generated in this way will be a good approximation provided the screened-ion potential is weakly scattering as it is for alkali ions, aluminum and some other metallic ions.

This charge density in the case of Aℓ agrees favourably with that obtained in a bandstructure calculation.[43]

The potential energy of the proton is now constructed with the UDA energy curve, the electrostatic correction (46) and the gradient corrections in (47). The results are presented in Table III for the proton situated at the octahedral and tetrahedral sites. As can be seen, the electrostatic correction, which is a Madelung type energy, is a very important contribution. The equilibrium site according to this calculation is the octahedral site and the energy at this position gives a heat of solution of 0.6 ev in fair agreement with experiment. The equilibrium site we find conflicts with some of the calculations using pseudopotentials however the latter are known to be sensitive to the choice of pseudopotential. The quasiatom approach retains many of the simplifying features of the pseudopotential theory but should be more reliable in its more realistic treatment of the metallic host.

We have outlined just some of the applications of the quasiatom concept. A systematic study of other combinations of impurities and hosts is necessary to establish the reliability of the approach and hence the degree of confidence that can be attached to it. On the basis of completed work the approach appears quite promising and certainly warrants further investigation.

IV Linear Response Theory

A. Formalism

Many problems arise in which the response of an electronic system to external perturbations is required. We encountered one specialized example of this when considering the response of the quasiatom to a nonuniform electronic background. Examples of more general interest include the response of atoms to various external perturbations, such as electric fields, field gradients, nuclear quadrupole moments, etc. or the response of a metallic system to applied magnetic fields. All of these represent a class of problems under the purview of linear response theory which is nothing more than (time-dependent) perturbation theory expressed in the modern terms of response and correlation functions. An excellent review of this general topic is given in Ref. 44.

The examples sited above are situations in which the external perturbation is time-independent. The resulting problem concerns the ground state properties in the presence of the perturbation and can therefore be safely approached using DFT. Recently, a number of workers[11,45,46] have independently applied DFT to the calculation of static dipole polarizabilities of closed shell atoms and ions. The results of these calculations will be reviewed in the following. An important generalization was made by Zangwill and Soven[46] to time-dependent phenomena without, however, any formal justification. Nonetheless, their successful calculation of the photoabsorption cross-section of rare gas atoms is unlikely to be accidental and offers some hope that a rigorous generalization of DFT to time-dependent phenomena may be possible. Much remains to be done to clarify this point.

In setting up the linear response formalism, we shall consider a slightly more general situation than described in the previous section by allowing the external perturbation, $v^{\sigma}_{ext}(\vec{r})$, to be spin-dependent, as indicated by the spin index $\sigma=\pm 1$. A concrete example is the Zeeman energy $\mathcal{H}_Z=-\sum_i \vec{\mu}_i.\vec{B}(\vec{r}_i)$ which accounts for the interaction of the electronic magnetic moments with an external magnetic field. The spin densities, $n^\sigma(\vec{r})$, for up- and down-spin electrons are now required to characterize the state of the system and the energy functional will depend on both. Equivalently, the independent variables can be chosen to be the particle number and magnetization densities,

$$n(\vec{r}) = \sum_{\sigma=\pm1} n^\sigma(\vec{r}) \tag{68}$$

and

$$m(r) = \sum_{\sigma=\pm1} \sigma n^\sigma(\vec{r}), \tag{69}$$

respectively. Kohn and Sham[2] were the first to suggest this spin generalization of DFT which was then developed further by other workers.

As a result of the external perturbation, the induced spin density is given, following our earlier argument for the response of the independent particle reference system, by

$$\delta n^\sigma(\vec{r}) = \int d\vec{r}' \chi^o_\sigma(\vec{r},\vec{r}')\delta v^\sigma_{eff}(\vec{r}') \tag{70}$$

where

$$\delta v^\sigma_{eff}(\vec{r}) = v^\sigma_{ext}(\vec{r}) + \delta\phi(r) + \delta v^\sigma_{xc}(\vec{r}). \tag{71}$$

For a paramagnetic state, the spin response function is one half of the density response function, $\chi^o_\sigma(\vec{r},\vec{r}') = \frac{1}{2}\chi^o(\vec{r},\vec{r}')$. The induced exchange-correlation potential is given to lowest order in the perturbation by

$$\delta v_{xc}^\sigma(\vec{r}) = \int d\vec{r}' [v_{xc}'(\vec{r},\vec{r}')\delta n(\vec{r}') + \sigma w_{xc}(\vec{r},\vec{r}')\delta m(\vec{r})] \tag{72}$$

where

$$v_{xc}'(r,r') = \left.\frac{\delta^2 E_{xc}[m,n]}{\delta n(\vec{r})\delta n(\vec{r}')}\right|_{\substack{n=n_o\\ m=o}} \tag{73}$$

and

$$\sigma w_{xc}(\vec{r},\vec{r}') = \left.\frac{\delta^2 E_{xc}[m,n]}{\delta m(\vec{r})\delta m(\vec{r}')}\right|_{\substack{n=n_o\\ m=o}} \tag{74}$$

In obtaining (73) and (74) we have used the fact that the expansion of $E_{xc}[m,n]$ in powers of m does not contain a linear term.

Using these results and taking the appropriate spin traces of (70), we find

$$\delta n(\vec{r}) = \int d\vec{r}'\chi^o(\vec{r},\vec{r}')[v_{ext}(\vec{r}') + \delta\phi(\vec{r}') + \int d\vec{r}'' v_{xc}'(\vec{r}',\vec{r}'')\delta n(\vec{r}'')] \tag{75}$$

and

$$\delta m(\vec{r}) = \int d\vec{r}'\chi^o(\vec{r},\vec{r}')[w_{ext}(\vec{r}') + \int d\vec{r}'' w_{xc}(\vec{r}',\vec{r}'')\delta m(\vec{r}'')] \tag{76}$$

with $v_{ext} = \sum_\sigma v^\sigma_{ext}$ and $w_{ext} = \sum_\sigma \sigma\, v^\sigma_{ext}$. We observe that the induced number and magnetization densities are decoupled, each satisfying its own integral equation. To lowest order, a magnetic field will not induce a charge density in a paramagnetic system and, conversely, an electrostatic perturbation will not induce a spin density. In the LDA, v_{xc} and w_{xc} both have the form shown in (64).

There are two stumbling blocks in applying (75) and (76). Firstly, we must be able to construct the response function $\chi^o(\vec{r},\vec{r}')$ which, for an inhomogeneous system, is a function of two spatial variables; and secondly, we must be able to solve a rather complicated integral equation for the induced densities. A formal prescription for obtaining χ^o is given below but it must be admitted that it can be determined explicitly only in favourable situations. The solution of the integral equation must rely on one's mathematical ingenuity!

A useful expression for $\chi^o(\vec{r},\vec{r}')$ can be obtained by considering the retarded single-particle Green's function defined by

$$\tilde{G}^+(\vec{r},\vec{r}',\omega) = \sum_i \frac{\psi_i(\vec{r})\psi_i^*(\vec{r}')}{\omega + i\eta - \varepsilon_i} \tag{77}$$

Here $\tilde{\psi}_i(\vec{r})$ is the wavefunction in the presence of the perturbation, $v_{ext}(\vec{r})$. The density is given in terms of $\tilde{G}^+$ by

$$n(\vec{r}) = -2\text{Im}\int_{-\infty}^{\varepsilon_F} \frac{d\omega}{\pi}\, \tilde{G}^+(\vec{r},\vec{r},\omega), \tag{78}$$

where the integration extends over the range of occupied states below the Fermi energy ε_F. It encompasses both bound and continuum states, depending on the choice of ε_F. The induced density is then given by the variation of (78),

$$\delta n(\vec{r}) = \delta[-2\text{Im}\int_{-\infty}^{\varepsilon_F} \frac{d\omega}{\pi}\, \tilde{G}^+(\vec{r},\vec{r},\omega)]. \tag{79}$$

But the Green's function has the perturbative expansion (in symbollic notation)

$$\tilde{G}^+ = G^+ + G^+ v_{ext} G^+ + \ldots \tag{80}$$

where G^+ is the Green's function <u>without</u> the perturbation. Using (80) in (79) we find (assuming a fixed chemical potential)

$$\delta n(\vec{r}) = -2\text{Im}\int_{-\infty}^{\varepsilon_F} \frac{d\omega}{\pi} \int d\vec{r}\,' G^+(\vec{r},\vec{r}\,',\omega) G^+(\vec{r}\,',r,\omega) v_{ext}(\vec{r}\,') \tag{81}$$

which immediately gives

$$\chi^o(\vec{r},\vec{r}\,') = -2\text{Im}\int_{-\infty}^{\varepsilon_F} \frac{d\omega}{\pi}\, G^+(\vec{r},\vec{r}\,',\omega) G^+(\vec{r}\,',\vec{r},\omega). \tag{82}$$

This expression is completely equivalent to the more-commonly-seen spectral representation

$$\chi^o(\vec{r},\vec{r}\,') = 2 \sum_{i,j} \psi_i(\vec{r})\psi_i^*(\vec{r}\,')\psi_j(\vec{r}\,')\psi_j^*(\vec{r})\, \frac{f(\varepsilon_i)-f(\varepsilon_j)}{\varepsilon_i-\varepsilon_j} \quad . \tag{83}$$

The main advantage of (82) over (83) is that the Green's function can sometimes be calculated directly, which obviates the need for explicitly performing the summation over all occupied <u>and</u> unoccupied single particle states. The equation determining the Green's function is

$$[-\tfrac{1}{2}\nabla^2 + v_{eff}(\vec{r})-\omega]G^+(\vec{r},\vec{r}\,',\omega) = -\delta(\vec{r}-\vec{r}\,') \tag{84}$$

and must be solved with the appropriate outgoing wave boundary condition. It should be noted that G^+ is a symmetric function of $\vec{r}$ and $\vec{r}\,'$.

B. The Spherical Cow

One situation in which the above theory can be applied effectively is when the unperturbed system is spherically symmetric. Some systems such as closed-shell atoms do possess this symmetry but even a representative atom in a metallic solid can be considered at a spherical entity for the purpose of calculating certain physical quantities. Some examples of the latter will be given later. Spherically symmetric situations are of course a theorist's delight and it is not unusual to find him shaping his problems accordingly.

If the unperturbed system is spherically symmetric, $\chi^o(\vec{r},\vec{r}')$ is only a function of the radial variables r,r' and the angle γ between the vectors $\vec{r}$ and $\vec{r}'$. It can therefore be expressed as

$$\chi^o(\vec{r},\vec{r}') = \sum_L \chi^o_\ell (r,r')Y_L(\hat{r})Y^*_L(\hat{r}') \tag{85}$$

where $Y_L(\vec{r})$ is a spherical harmonic and the index L stands for the quantum numbers ℓ and m. The angular components, $\chi^o_\ell(r,r')$, relate to the multipole moment of order 2^ℓ induced by the external perturbation.

Similarly we have the expression

$$G^+(\vec{r},\vec{r}';\omega) = \sum_L G_\ell^+(r,r',\omega)Y_L(\hat{r})Y^*_L(\hat{r}'). \tag{86}$$

Using the expansions (85) and (86) in (82) we obtain

$$\chi_\ell{}^o(r,r') = -4\pi \mathrm{Im} \int_{-\infty}^{\varepsilon_F} \frac{d\omega}{\pi} \sum_{\ell',\ell''} C_{\ell\ell'\ell''}G^+_{\ell'}(r,r;\omega)G^+_{\ell''}(r;r,\omega) \tag{87}$$

with

$$C_{\ell\ell'\ell''} = \left(\frac{2\ell'+1}{4\pi}\right)\left(\frac{2\ell''+1}{4\pi}\right)\int_{-1}^{1} dx P_\ell(x)P_{\ell'}(x)P_{\ell''}(x). \tag{00}$$

Here $P_\ell(x)$ is a Legendre polynomial. Some special cases of (87) which will be used in the following are:

$$\chi_0^o(r,r') = -\frac{1}{2\pi}\mathrm{Im}\int_{-\infty}^{\varepsilon_F} \frac{d\omega}{\pi} \sum_{\ell=o}^{\infty} (2\ell+1)G_\ell^+(r,r';\omega)G_\ell^+(r';r,\omega) \qquad (89)$$

$$\chi_1^o(r,r') = -\frac{1}{\pi}\mathrm{Im}\int_{-\infty}^{\varepsilon_F} \frac{d\omega}{\pi} \sum_{\ell=o}^{\infty} (\ell+1)G_\ell^+(r,r';\omega)G_{\ell+1}^+(r';r,\omega) \qquad (90)$$

$$\chi_2^o(r,r') = -\frac{1}{2\pi}\mathrm{Im}\int_{-\infty}^{\varepsilon_F} \frac{d\omega}{\pi} \sum_{\ell=o}^{\infty} \left\{\frac{\ell(\ell+1)(2\ell+1)}{(2\ell-1)(2\ell+3)} G_\ell^+(r,r';\omega)G_\ell^+(r';r,\omega)\right.$$

$$\left. + \frac{3(\ell+1)(\ell+2)}{2\ell+3} G_\ell^+(r,r';\omega)G_{\ell+2}^+(r';r,\omega)\right\} \qquad (91)$$

For these, and the higher angular momentum components, two kinds of products appear: $G_\ell^+G_\ell^+$, and $G_\ell^+G_{\ell'}^+$, with $\ell'\neq\ell$. The latter is simpler to deal with and will be taken up first.

Writing the normalized electronic states as

$$\psi_i(\vec{r}) = R_{n_i\ell_i}(r)Y_{L_i}(\hat{r}), \qquad (92)$$

we can obtain G_ℓ^+ from (77) as

$$G_\ell^+(r,r';\omega) = \sum_i \delta_{\ell,\ell_i} \frac{R_{n_i\ell_i}(r)R_{n_i\ell_i}(r')}{\omega-\varepsilon_{n_i\ell_i}+i\eta} \quad . \qquad (93)$$

This shows that

$$\mathrm{Im}\ G_\ell^+(r,r';\omega) = -\pi \sum_i \delta_{\ell,\ell_i} R_{n_i\ell_i}(r)R_{n_i\ell_i}(r')\delta(\omega-\varepsilon_{n_i\ell_i}) \qquad (94)$$

which can be used to perform the frequency integration in (90). Alternatively, $G_\ell^+(r,r',\omega) \equiv g_\ell(r,r',\omega)/rr'$ can be obtained as the solution of

$$\left[-\frac{1}{2}\frac{d^2}{dr^2} + \frac{\ell(\ell+1)}{2r^2} + v_{eff}(r) - \omega\right] g_\ell(r,r';\omega) = -\delta(r-r'). \qquad (95)$$

g_ℓ is readily expressible in terms of the linearly independent solutions to the homogeneous radial equation at the energy $\omega = \frac{1}{2}k^2$:

$$\left[-\frac{d^2}{dr^2} + \frac{\ell(\ell+1)}{r^2} + 2v_{eff}(r)-k^2\right]u_{\ell k}(r) = o. \qquad (96)$$

These solutions can be chosen such that the first, $\phi_{\ell k}(r)$, is regular at the origin while the second, $\chi_{\ell k}^{(1)}(r)$, behaves asymptotically for $r\to\infty$ as $rh_\ell^{(1)}(kr)$ ($h_\ell^{(1)}$, a Hankel function). In terms of these functions,

$$G_\ell^+(r,r,'\omega) = \frac{2}{W[\phi_{\ell k},\chi_{\ell k}^{(1)}]} \; \frac{\phi_{\ell k}(r_<)\chi_{\ell k}^{(1)}(r_>)}{rr'} \qquad (97)$$

Here $r_<(r_>)$ is the lesser (greater) of r and r' and $W[\phi_{\ell k},\chi_{\ell k}^{(1)}]$ is the Wronskian of the two solutions,

$$W[\phi_{\ell k},\chi_{\ell k}^{(1)}] = \phi_{\ell k} \frac{d\chi_{\ell k}^{(1)}}{dr} - \frac{d\phi_{\ell k}}{dr} \chi_{\ell k}^{(1)} \; . \qquad (98)$$

If ω happens to be at a bound state energy, $\phi_{\ell k}$ and $\chi_{\ell k}$ as defined are linearly dependent and hence the Wronskian vanishes. This singular property of the Green's function is of course apparent from the spectral representation (93).

If ω does not correspond to a bound state energy (97) can be simplified further by normalizing $\phi_{\ell k}(r)$ so that it behave for $r\to\infty$ as $r[\gamma h_\ell^{(1)}(kr)+h_\ell^{(2)}(kr)]$. With this choice, (97) becomes

$$G_\ell^{\,+}(r,r';\omega) = -ikR_\ell(r_<;k)R_\ell^{(1)}(r_>;k) \qquad (99)$$

where $R_\ell(r;k) = \phi_{\ell k}(r)/r$ and $R_\ell^{(1)}(r;k) = \chi_{\ell k}^{(1)}(r)/r$. For $\omega<0$,k is purely imaginary ($k=i\kappa$) and $G_\ell^+(r,r',\omega)$ is real.

In evaluating $\mathrm{Im}[G_\ell^+G_{\ell'}^+]$ with $\ell'\neq\ell$ in the range $\omega<0$, we can use (94) for the imaginary part and (99) for the real part. The frequency integral then picks off the various bound state contributions and we then obtain for the case $\ell=1$,

$$\chi_1^o(r,r') = \frac{1}{\pi}\sum_{k,occ}(-1)^{\ell_i+1}R_{n_i\ell_i}(r)R_{n_i\ell_i}(r')[\ell_i\tilde{R}_{\ell_i-1}(r_<;\kappa_{n_i\ell_i})\tilde{R}^{(1)}_{\ell_i-1}(r_>;\kappa_{n_i\ell_i})$$

$$+ (\ell_i+1)\tilde{R}_{\ell_i+1}(r_<;\kappa_{n_i\ell_i})\tilde{R}^{(1)}_{\ell_i+1}(r_>;\kappa_{n_i\ell_i})] \qquad (100)$$

where quantities with a tilde have been introduced according to $f_\ell(r;i\kappa)=i^{\ell}\tilde{f}_\ell(r;\kappa)$. Similar expressions are obtained in all cases in which the product $G_\ell^+G_{\ell'}^+$, with $\ell\neq\ell'$ appears. The summation in (100) includes all occupied bound states and is the expression relevant to determining the dipole response of an atom or ion. The labour required to construct $\chi_1^{\,o}$ increases in proportion to the number of occupied states and therefore poses no problems even for heavy atoms. Information concerning the unoccupied spectrum is contained in the non-physical radial solutions $R_\ell(r;k)$ and $R_\ell^{(1)}(r;k)$.

The response of an electron gas will also include a contribution from continuum states with $\omega>0$. In this case it is convenient to choose a different pair of linearly independent solutions which behave for $r\to\infty$ as

$$u_{\ell k}(r) \to r[\cos\delta_\ell(k)j_\ell(kr) - \sin\delta_\ell(k)n_\ell(kr)] \tag{101}$$

and

$$v_{\ell k}(r) \to r[\cos\delta_\ell(k)n_\ell(kr) + \sin\delta_\ell(k)j_\ell(kr)] \tag{102}$$

where $\delta_\ell(k)$ are the scattered-wave phase shifts for the potential $v_{eff}(r)$. $u_{\ell k}(r)$ is in fact the same as $\phi_{\ell k}(r)$ while

$$\chi_{\ell k}^{(1)}(r) = e^{-i\delta_\ell(k)}[u_{\ell k}(r) + i v_{\ell k}(r)] \tag{103}$$

In terms of these functions ($\omega>0$),

$$G_\ell^+(r,r',\omega)G_{\ell'}^+(r',r,\omega) = -4k^2 u_{\ell k}(r_<)u_{\ell' k}(r_<)[u_{\ell k}(r_>) + i v_{\ell k}(r_>)] \times [u_{\ell' k}(r_>) + i v_{\ell' k}(r_>)] \tag{104}$$

This result is valid for any value of ℓ and ℓ'. As an example, the continuum part of $\chi_o^o(r,r')$ is given by

$$\chi_o^{oc}(r,r') = \frac{4}{\pi^2}\int_0^{k_F} dk k^3 \sum_{\ell=0}^{\infty} (2\ell+1)u_{\ell k}^2(r_<)u_{\ell k}(r_>)v_{\ell k}(r_>). \tag{105}$$

In this case the response function calculation is reduced to a simple quadrature. The Knight shift calculation to be described later makes use of (105).

So far we have managed to skirt around the problem associated with evaluating $\mathrm{Im}G_\ell^+G_\ell^+$ for negative frequencies. The difficulty arises because of the double poles occurring in this product at each of the bound state energies. Handling the double pole requires a careful treatment of the singularities in the Green's function.

Let us write (97) as

$$G^+(E) = \frac{N(E)}{W(E)} \tag{106}$$

where the numerator is

$$N(E) = 2\phi_{\ell k}(r_<)\chi_{\ell k}^{(1)}(r_>)/rr'. \tag{107}$$

Here we have suppressed the dependence of G^+ and N on r,r' and ℓ, exhibiting only its energy dependence. Near a bound state energy, E_o, N and W can be expanded as

$$N(E) = N(E_o) + N^{(1)}(E_o)(E-E_o) + \ldots \tag{108}$$

and

$$W(E) = W^{(1)}(E_o)(E-E_o) + \tfrac{1}{2}W^{(2)}(E_o)(E-E_o)^2 + \ldots \tag{109}$$

where we have used the fact that the Wronskian vanishes at $E=E_o$. Using these expansions we can separate $G^+(E)$ into singular and non-singular parts in the vicinity of E_o:

$$G^+(E) = G^+(E)_{\text{sing}} + G^+(E)_{\text{non-sing}} \tag{110}$$

with

$$G^+(E) = \frac{N(E_o)}{W^{(1)}(E_o)(E-E_o)} = \frac{R_o(r)R_o(r')}{E-E_o} \tag{111}$$

and

$$G^+(E_o)_{\text{non-sing}} = \frac{1}{W^{(1)}(E_o)}\left[N^{(1)}(E_o) - N(E_o)W^{(2)}(E_o)/W^{(1)}(E_o)\right] \tag{112}$$

We note that $G^+(E_o)_{non-sing}$ is real at the negative energies of interest.

With these results, we find that

$$\mathrm{Im}[G^+(E)G^+(E)] = \pi R_o^2(r)R_o^2(r') \frac{\partial}{\partial E}\,\delta(E-E_o) - 2\pi\delta(E-E_o)R_o(r)R_o(r')G^+(r,r',E_o)_{non-sing} \tag{113}$$

with a similar contribution occurring for each bound state. The first term in (113) vanishes upon integrating over ω in (89) - (91). Thus, for example, the bound state part of $\chi_o^o\,(r,r')$ is given by

$$\chi_o^{oB}(r,r') = \frac{1}{\pi}\sum_{i,occ}\;(2\ell_i+1)R_{n_i\ell_i}(r)R_{n_i\ell_i}(r')G^+_{\ell_i}(r,r',\kappa_{n_i\ell_i})_{non-sing} \tag{114}$$

The explicit evaluation of $G^+_{non-sing}$ is lengthy and will not be given here. We refer to Ref. 49 for details.

C. Applications

1. Dipole Polarizabilities[11]

The application of a uniform electric field $\hat{E}$ of unit amplitude corresponds to the external potential

$$v_{ext}(\vec{r}) = \hat{E}.\vec{r} = \frac{4\pi}{3}\,r\sum_{m=-1}^{1}Y_{1m}(\hat{r})Y^*_{1m}(\hat{E}) \tag{115}$$

The induced density for a spherical atom can be written as

$$\delta n(\vec{r}) = -\,\alpha(r)\hat{r}.\hat{E} = -\,\frac{4\pi}{3}\alpha(r)\sum_{m=-1}^{1}Y_{1m}(\hat{r})Y^*_{1m}(\hat{E}) \tag{116}$$

which isolates the interesting radial dependence in the function $\alpha(r)$. Substituting (115) and (116) into (75) we obtain an integral equation for $\alpha(r)$:

$$\alpha(r) = -\int_o^\infty dr'r'^3\chi_1^o(r,r') + \int_o^\infty dr'r'^2\chi_1^o(r,r')\int_o^\infty dr''r''^2[\gamma_1(r',r'') + v'_{xc,1}(r',r'')] \times\alpha(r'') \tag{117}$$

where

$$\gamma_\ell(r,r') = 4\pi/(2\ell+1)r_<^\ell/r_>^{\ell+1} \tag{118}$$

and

$$v'_{xc,\ell}(r,r') = \int d\hat{r}\int d\hat{r}'Y^*_L(\hat{r})v'_{xc}(\vec{r},\vec{r}')Y_L(\hat{r}') \tag{119}$$

The response function $\chi_1^{o}(r,r')$ is given in (100).

The dipole moment of the atom is given by

$$\vec{p} = -\int d\vec{r}\vec{r}\delta n(\vec{r}) \tag{120}$$

and the polarizability, defined by $\vec{p} = \alpha\vec{E}$, is obtained from the integral

$$\alpha = \frac{4\pi}{3}\int_0^\infty drr^3\alpha(r) \tag{121}$$

The solution of (117) was obtained by using Simpson's rule to reduce the integral equation to a system of linear equations determining $\alpha(r)$ at a discrete set of points $\{r_j\}$. In doing so, care must be taken to account for the cusp in $\chi_1^o(r,r')$ at $r=r'$. Because of the large 'diagonal' component on the left hand side of (117), the equations could be solved efficiently using Gaussian elimination. Typically, the dimensionality of the problem ranged between 50 and 120. The numerical accuracy of the calculated polarizabilities is estimated to be better than one part in 10^3.

A calculation of the HF polarizability of He using the above method gave the value $\alpha=1.3223$ a.u. which is in excellent agreement with other calculations. This calculation served as a check on the calculation of $\chi_1^o(r,r')$ and the subsequent solution of the integral equation.

The DFT results were obtained in the LDA. Self-consistent atomic calculations for each of the atoms were performed, and the resulting effective potential and density were used in the evaluation of $\chi_1^o(r,r')$ and $v'_{xc}(r,r')$. The radial dipole distributions for the rare gases are illustrated in Fig. 7 and the corresponding polarizabilities are presented in Table IV. For comparison, HF, CI and experimental values are also shown, where available. It is apparent that the DFT results are in good agreement with experiment, except for He and Ne. The agreement for the alkali ions and alkaline earth atoms is also very good, rivalling even the most sophisticated CI calculations for the heavier atoms.

Also shown in Table IV are results denoted by α_o which represent the polarizability of the atom neglecting the interactions between the electrons, i.e., omitting the second term on the right hand side of (117). These results are analogous to uncoupled -HF in that the distortion of the single particle orbitals is calculated without including the change in the self-consistent potential due to the electronic interactions. It is clear that self-consistency plays an important role. Physically it accounts for the screening of the externally applied field by the field due to the induced dipole moment. This effect is illustrated for Be in Fig. 8 which shows the radial dependence of the effective potential $\delta v_{eff}(r)$ as compared to the linear dependence of $v_{ext}(r)$. The reduction in $\delta v_{eff}(r)$ is due to the shielding provided by the induced density on the periphery of the atom. In fact the external field is overscreened in the core of the atom so that the 1s orbital is polarized opposite to the 2s orbital.

It is apparent that the worst results are those for the lightest atoms He, Li^+, Be and Ne. The fact that the polarizabilities are overestimated suggests that these atoms are more tightly bound than predicted by the LDA. This is indeed the case as can be seen readily for the example of He. It is found that the HF potential for He confines the electrons more than the LDA potential, resulting in the much smaller HF polarizability. Furthermore, the HF result is rather close to CI and so the failure of the LDA is not related to correlation effects. Rather it must be attributed to the inadequacy of local exchange in a two electron atom. For the heavier atoms statistical exchange seems to be a better approximation and with the inclusion of correlation the results are in fact superior to HF.

As discussed at the end of Sec. IIB the exchange term in HF theory cancels the self-interaction. The failure to do so in the LDA can lead to problems as observed here for the polarizabilities of light atoms. Since self-interaction corrections have been proposed,[17-19] albeit somewhat arbitrarily, it was of some interest to see whether the LDA results could be improved. Introducing these corrections into the polarizability calculations is possible but because each orbital now has its own potential the simplicity of the original formulation in terms of the total density is lost. Eq. (75) must be generalized to a set of coupled integral equations for each of the orbital densities. Nonetheless the problems are not insurmountable and SIC polarizabilities were calculated for some selected atoms. These results are also displayed in Table IV.

It cannot be said that the SIC results are a marked improvement over LDA. As expected, the polarizability of He is reduced considerably, but to the extent that it now underestimates the correct value. Evidently the self-interaction corrections have gone too far in making the atom less polarizable. This is especially clear in the case of the negative hydrogen ion. In the LDA H^- is not bound whereas with SIC it is, having a ground state energy of -1.039 Ry in reasonable agreement with the known electron affinity of H. However the calculated polarizability is a gross underestimate of the accurate variational value ($206.4a_o^3$).[50] Although the SIC have many desirable features they are still not totally adequate.

2. Quadrupole Polarizabilities and Sternheimer Antishielding Factors.

We next consider the response of an atom to an $\ell=2$ perturbation such as would arise for an imposed electric field gradient or from the quadrupole moment of a nucleus. This problem is most relevant in the study of pure and impure metals using the technique of quadrupole resonance. In such experiments one probes the electric field gradient at the resonant nucleus produced by the total electron charge density in the solid. The distortion of the ion core itself contributes to the field gradient and can be accounted for approximately by considering the response of the free ion.[51]

An axially symmetric field gradient is represented by the potential

$$v_{ext}(\vec{r}) = -\frac{1}{2}\, qr^2 P_2(\hat{r}.\hat{I})$$

$$= -\frac{2\pi}{5}\, qr^2 \sum_{m=-2}^{2} Y_{2m}^{*}(\hat{I})Y_{2m}(\hat{r}) \qquad (122)$$

where q is the principle value of the electric field gradient tensor along the principle axis $\hat{I}$. The electron density induced by this perturbation can be expressed as

$$\delta n(\vec{r}) = 4\pi\alpha_q(r) \sum_{m=-2}^{2} Y^*_{2m}(\hat{I})Y_{2m}(\hat{r}) \tag{123}$$

This change in density produces an electric field gradient at the position of the nucleus given by

$$q' = \int d\vec{r}[\tfrac{2}{r^3} P_2(\hat{r}.\hat{I})]\delta n(\vec{r})$$

$$= 8\pi \int_0^\infty dr\alpha_q(r)/r. \tag{124}$$

The total field gradient, q+q', is the quantity determining the magnitude of the quadrupole splittings. The screening of the external field is conventionally represented by the ratio γ_∞=q'/q, the Sternheimer antishielding factor, and the total field is written as

$$q_{total} = (1+\gamma_\infty)q. \tag{125}$$

The field gradient also induces a quadrupole moment in the electron distribution which defines the quadrupole polarizability according to

$$\alpha_q = \int d\vec{r} \, [-2r^2P_2(\hat{r}.\hat{I})]\delta n(\vec{r})/q$$

$$= -8\pi/q \int_0^\infty drr^4\alpha_q(r). \tag{126}$$

This quantity is essentially of theoretical interest since it is not experimentally available.

The procedure for obtaining $\alpha_q(r)$ is identical to the case of the dipole response apart from $\chi_2^o(r,r')$, given by (91), appearing in the integral equation. The calculation of χ_2^o is somewhat more involved since there are two contributions coming from $G^+_\ell G_\ell^+$ and $G^+_\ell G^\pm_{\ell+2}$. The latter can be expressed in the form of (100); it accounts for the "angular excitations" leading to shielding of the external field.[52]

The $G_\ell^+ G_\ell^+$ contribution is similar to the result shown in (114); it contains the "radial excitations" which give rise to anti-shielding. In atoms with non-s-state orbitals the radial excitations tend to dominate the angular excitations.

Results for selected ions are given in Table V . A comparison of the polarizabilities with (α_q) and without ($\alpha_q^{\,o}$) interactions reveals that self-consistency is far less important for $\ell=2$ perturbations. The externally applied potential, proportional to r^2, tends to dominate the induced potential in most regions of the ion.

This is also true of the antishielding factor for those ions with a rare gas structure. For the noble metal ions however self-consistency is much more important, particularly for γ_∞ which is reduced significantly from $\gamma_\infty^{\,o}$.

Also shown in Table V for comparison are the HF results of Feiock and Johnson.[53] These are only a representative sampling of the vast number of calculations performed but are commonly quoted in the literature since they provide a comprehensive tabulation based on one theoretical scheme. The LDA results for γ_∞ agree well with those of FJ in most cases in which self-consistency is of secondary importance. Two exceptions to this are Cs^+ and Ba^{2+} for which the relativistic corrections included by FJ are probably important. The LDA calculations are of course non-relativistic. The LDA polarizabilities for the alkali and alkaline earth ions consistently lie slightly above the HF values.

The most interesting comparison occurs for the noble metal ions for which, as we have already emphasized, self-consistency is important. Our values of the independent particle shielding factors $\gamma_\infty^{\,o}$ are in fact quite close to those of FJ which, bearing in mind the agreement for the other ions, is to be expected since their calculations are uncoupled-HF. The reduction of γ_∞ from $\gamma_\infty^{\,o}$ is clearly a real effect which happens to be particularly evident for the noble metal ions.

Similar results have recently been obtained by G.D. Mahan[54] using an extension of the Sternheimer approach which includes the effects of self-consistency as described above. In addition, Mahan has gone further and has calculated octapole and hexadecapole polarizabilities. DFT lends itself nicely to calculations of this kind, allowing the important effects of self-consistency to be included in a straightforward and physically transparent way.

3. Knight Shifts

As a final application of the linear response theory we shall describe some recent calculations of the Knight shift of host nuclei and of muons in simple metals.[48] The Knight shift is defined as the fractional shift of the magnetic resonance frequency from its value in the externally applied field B_o. The shift is due to the interaction of the nucleus with the electrons in the metal and has a contribution from the dipolar fields of the electronic spins as well as from the fields due to their orbital motion (diamagnetic currents). Usually the dipolar field is the dominant contribution. Because of the overlap of the electronic wavefunction with the nucleus and the singular nature of the dipolar field, the final interaction takes the form of a contact term which is proportional to the electron magnetization at the nuclear spin.[55] The Knight shift is then given by

$$K_s = \frac{8\pi}{3} \frac{m(o)}{B_o} \tag{127}$$

where m(o) is the electron-spin magnetic-moment density at the position of the nucleus. Simply stated, the problem consists of evaluating m(o) when a magnetic field B_o is applied to the metal.

If the electrons in the metal are taken to be non-interacting, that is the electronic wavefunction is approximated as a single Slater determinant, the effect of the magnetic field is to redistribute the occupation of the electronic band states at the Fermi level. In this approximation, the Knight shift can be written as[55]

$$K_s = \frac{8\pi}{3} \chi_s <|\psi(o)|^2>_{E_F} \tag{128}$$

where the angular brackets denote an average of the contact density $|\psi(o)|^2$ over the Fermi surface. The spin susceptibility χ_s is strictly speaking the bandstructure susceptibility of the metal but the effect of exchange interactions is often included by using the exchange-enhanced susceptibility as determined experimentally (or perhaps theoretically).

In terms of DFT the approximation (128) corresponds to neglecting the exchange-correlation field appearing in (76), or more realistically, to replacing $w_{xc}(\vec{r},\vec{r}')$ by a spatial average. In this case,

$$m(\vec{r}) \approx \int d\vec{r}'\chi^{o}(\vec{r},\vec{r}')B_{eff} \tag{129}$$

where $B_{eff} = B_o + B_{xc}$ is an effective magnetic field acting on the electronic spins. Since the solid is highly inhomogeneous, the effective field is actually spatially dependent

$$B_{eff}(\vec{r}) = B_o + \int d\vec{r}' w_{xc}(\vec{r},\vec{r}')m(\vec{r}') \tag{130}$$

and $m(\vec{r})$ must be obtained as the solution of an integral equation (or, alternatively, from a self-consistent spin-polarized bandstructure calculation). The physical implication of a spatially varying field is that states below the Fermi level, including the core states, contribute to m(o) in contrast to (128). This effect is referred to as exchange core polarization.[56]

We have based our Knight shift calculations on the integral equation (76) with $\chi^{o}(\vec{r},\vec{r}')$ given by (82). However we have made one important approximation which simplifies the analysis considerably. The Green's function appearing in (82) is for a periodic solid, and its evaluation is as difficult as doing a full bandstructure calculation. Since we are interested in obtaining m(o) at the position of one particular ion, it is the local electronic properties which are most important. For this reason we have considered a model in which the ion of interest resides at the centre of a vacancy in an otherwise homogeneous electron gas. The vacancy simulates the actual situation in the metal but the replacement of all other ions by a uniform gas must be justified. It is in fact a poor approximation if each ionic potential is a strong scatterer, resulting in a bandstructure which is far from free-electron like. For many of the simple metals this is not the case and the single-ion approximation is a useful simplification.

To proceed we must first determine the self-consistent potential and density of the single ion as described in Sec. IIIC. With this information, $\chi_0^o(r,r')$ is evaluated using (105) and (114). Since the perturbation and induced magnetization are not localized quantities but tend to finite, constant values far from the ion it is not possible to use (105) to describe the electron gas response since an infinite number of ℓ-values in the summation would be required. We note however, that far from the ion the response is that of a uniform electron gas. If we then consider the quantity

$$\Delta\chi^o(\vec{r},\vec{r}') = \chi^o(\vec{r},\vec{r}') - \overline{\chi}^o(\vec{r},\vec{r}') \tag{131}$$

where $\overline{\chi}^o$ is the response function for a homogeneous system, the difference $\Delta\chi^o$ will be a localized response function and can be evaluated by retaining a finite number of ℓ-values (typically seven). Similarly, it is convenient to define the localized magnetization

$$\Delta m(\vec{r}) = m(\vec{r}) - \overline{m} \tag{132}$$

where $\overline{m}$ is the magnetization at large distances. An integral equation for $\Delta m(\vec{r})$ can be formulated and solved as in the polarizability calculations.

Results for a number of simple metals are presented in Table VI. The Knight shifts calculated in this way are in reasonable agreement with experiment, which provides some *a posteriori* justification of the approximations made. The values for Li, Na, K and Mg are in fact close to those obtained recently using a spin-polarized band structure calculation.[57] The large discrepancy for Be is a consequence of the strength of the Be potential which results in the Fermi surface density of states being much smaller than in the free electron approximation. The poor results for Rb and Cs are probably due to the neglect of relativistic effects.

The total Knight shift has been divided into a core and valence contribution in Table VI. The former is typically 10% of the latter, and of opposite sign. The negative core polarization is due entirely to the inhomogeneous effective field acting on the core states; a uniform field would not induce any polarization. Even though the core polarization appears relatively small, its self-consistent contribution to the effective field has an important effect. The Knight shift calculated with the uniform field approximation (128) is typically 20% larger than the self-consistent results.

Recently, the Knight shift of muons in metals has been measured using the muon spin rotation technique.[58] Since the muon resides at an interstitial site, a model in which the muon is screened by a uniform electron gas has been used to interpret the results.[59,39] Calculations of the muon Knight shift as a function of the gas density have been done using both a direct spin-polarized approach and the linear response theory described above. The results obtained by either method are in substantial agreement and are compared with experiment in Fig. 9. In contrast to the host Knight shifts, large discrepancies - even in sign - are the rule. The possible importance of diamagnetic shielding has been proposed[60] but recent calculations indicate this not to be the case.[61] Clearly the main limitation of the model is the neglect of the crystal structure and more refined calculations taking this into account will probably be necessary to resolve the existing discrepancies.

Acknowledgements

Needless to say, the work described here was very much a group effort. It is a pleasure to give credit to my congenial collaborators, Drs. M.J. Stott, D. Zobin and T. McMullen. Their permission to present unpublished material is greatly appreciated.

This work was supported by a grant from the Natural Sciences and Engineering Research Council, Canada.

References

1. P. Hohenberg and W. Kohn, Phys. Rev. 136, B864 (1964).
2. W. Kohn and L.J. Sham, Phys. Rev. 140, A1133 (1965).
3. W.F. Saam and C. Ebner, Phys. Rev. A15, 2566 (1977).
4. X. Campi and D.W. Sprung, Nuc. Phys. A194, 401 (1972).
5. L.H. Thomas, Proc. Camb. Philos. Soc. 23, 542 (1926); E. Fermi, R.c. Acc. Lincei 6, 602 (1927).
6. J.C. Slater, Phys. Rev. 81, 385 (1951).
7. N.D. Lang, Solid State Physics, ed. by H. Ehrenreich, F. Seitz and D. Turnbull (Academic Press, New York, 1973), Vol. 28, p. 225.
8. O. Gunnarsson, in Electrons in Disordered Metals and at Metallic Surfaces, ed. by P. Phariseau, B.L. Györffy and L. Scheire (Plenum, New York, 1979), p. 1.
9. M.J. Stott and E. Zaremba, Solid State Commun. 32, 1297 (1979).
10. M.J. Stott and E. Zaremba, Phys. Rev. B22, 1564 (1980).
11. M.J. Stott and E. Zaremba, Phys. Rev. A21, 12 (1980); Erratum, Phys. Rev. A 22, 2293 (1980)
12. D.J.W. Geldart and M. Rasolt, Phys. Rev. B13, 1477 (1976).
13. J.P. Perdew, D.C. Langreth and V. Sahni, Phys. Rev. Lett. 38, 1030 (1977).
14. O. Gunnarsson and B. I. Lundqvist, Phys. Rev. B13, 4274 (1976).
15. H.B. Shore, J.H. Rose and E. Zaremba, Phys. Rev. B15, 2858 (1977).
16. K. Schwarz, Chem. Phys. Lett. 57, 605 (1978); J. Phys. B11, 1339 (1978).
17. G.W. Bryant and G.D. Mahan, Phys. Rev. B17, 1744 (1978).
18. J.P. Perdew, Chem. Phys. Lett. 64, 127 (1979).
19. A. Zunger and M.L. Cohen, Phys. Rev. B18, 5449 (1978).
20. R.T. Sharp and G.K. Horton, Phys. Rev. 90, 317 (1955).
21. J.D. Talman and W.F. Shadwick, Phys. Rev. A14, 36 (1976); K. Aashamar, T.M. Luke and J.D. Talman, ibid. A19, 6 (1979); P.S. Ganas, J.D. Talman and A.E.S. Green, ibid. A22, 336 (1980).

22. O. Gunnarsson, M. Jonson and B.I. Lundqvist, Phys. Lett. 59A, 177 (1976); Solid State Commun. 24, 765 (1977); Phys. Rev. B20, 3136 (1979). J.A. Alonso and L.A. Girifalco, Phys. Rev. B17, 3735 (1978).

23. B.Y. Tong and L.J. Sham, Phys. Rev. 144, 1 (1966); O. Gunnarsson, J. Appl. Phys. 49, 1399 (1978).

24. O. Gunnarsson, J. Harris and R.O. Jones, Phys. Rev. B15, 3027 (1977).

25. J.F. Janak and A.R. Williams, Phys. Rev. B14, 4199 (1976); V.L. Moruzzi, A.R. Williams and J.F. Janak, Phys. Rev. B15, 2854 (1977).

26. H.D. Cohen, J. Chem. Phys. 43, 3558 (1965).

27. A. Dalgarno, Adv. Phys. 11, 281 (1962).

28. N.D. Lang and A.R. Williams, Phys. Rev. Lett. 34, 531 (1975); ibid. 37, 212 (1976); O. Gunnarsson, H. Hjelmberg and B.I. Lundqvist, Surf. Sci. 63, 348 (1977); H. Hjelmberg, O. Gunnarsson and B.I. Lundqvist, ibid. 68, 158 (1977).

29. H. Hjelmberg, Phys. Scr. 18, 481 (1978).

30. R.P. Messmer and D.R. Salahub, Phys. Rev. B16, 3415 (1977).

31. J.K. Nørskov and N.D. Lang, Phys. Rev. B21, 2131 (1980); D.S. Larsen and J.K. Nørskov, J. Phys. F9, 1975 (1979).

32. Z.D. Popović and M.J. Stott, Phys. Rev. Lett. 33, 1164 (1974); C.-O. Almbladh, U. von Barth, Z.D. Popović and M.J. Stott, Phys. Rev. B14, 2250 (1976); E. Zaremba, L.M. Sander, H.B. Shore and J.H. Rose, J. Phys. F7, 1763 (1977).

33. M.D. Whitmore, J. Phys. F6, 1259 (1976).

34. J.K. Nørskov, Solid State Commun. 24, 691 (1977).

35. M. Manninen, P. Hautojärvi and R. Nieminen, Solid State Commun. 23, 795 (1977).

36. M. Manninen, R. Nieminen, P. Hautojärvi and J. Arponen, Phys. Rev. B12, 4012 (1975).

37. H.F. Budd and J. Vannimenus, Phys. Rev. Lett. 31, 1218 (1973).

38. B.M. Deb, Rev. Mod. Phys. 45, 22 (1973).

39. M. Manninen and R.M. Nieminen, J. Phys. F 9, 1333 (1979).

40. D. Herlach, Second International Topical Meeting on Muon Spin Rotation, Vancouver, Canada 1980, to be published.

41. N.D. Lang and W. Kohn, Phys. Rev. B1, 4555 (1970).

42. T. McMullen, to be published.

43. J.P. Walter, C.Y. Fong and M.L. Cohen, Solid State Commun. 12, 303 (1973).

44. P.C. Martin, Many-Body Physics, Ed. C. deWitt and R. Balian (Gordon and Breach, New York, 1968), p. 37.

45. G.D. Mahan, Solid State Commun. 33, 797 (1980).

46. A. Zangwill and P. Soven, Phys. Rev. A21, 1561 (1980).

47. U. von Barth and L. Hedin, J. Phys. C5, 1629 (1972); A.K. Rajagopal and J. Callaway, Phys. Rev. B7, 1912 (1973); S.H. Vosko and J.P. Perdew, Can. J. Phys. 53, 1385 (1975).

48. E. Zaremba and D. Zobin, Phys. Rev. Lett. 44, 175 (1980).

49. M.J. Stott, E. Zaremba and D. Zobin, Can. J. Phys. 60, 210 (1982).

50. R.M. Glover and F. Weinhold, J. Chem. Phys. 65, 4913 (1976).

51. R.M. Sternheimer, Phys. Rev. 96, 951 (1954); 107, 1565 (1957); 115, 1198 (1959).

52. M.H. Cohen and F. Reif, Solid State Physics, ed. by F. Seitz and D. Turnbull (Academic Press, New York, 1957), vol. 5, p. 321.

53. F.D. Feiock and W.R. Johnson, Phys. Rev. 187, 39 (1969).

54. G.D. Mahan, Phys. Rev. A 22, 1780 (1980).

55. C.P. Slichter, Principles of Magnetic Resonance (Springer-Verlag, Berlin, 1978), 2nd ed.

56. S.D. Mahanti, L. Tterlikkis and T.P. Das, in Magnetic Resonance, edited by C.K. Coogan et al. (Plenum, New York, 1970).

57. L. Wilk and S.H. Vosko, Can. J. Phys. 59, 888 (1981).

58. M. Camani, F.N. Gygax, W. Rüegg, A. Schenck, H. Schilling, E. Klempt, R. Schulze and H. Wolf, Phys. Rev. Lett. 42, 679 (1979).

59. R. Munjal and K. Petzinger, Hyperfine Interact. 4, 301 (1978).

60. J. Keller and A. Schenck, Hyperfine Interact. 6, 39 (1979).

61. E. Zaremba and D. Zobin, Phys. Rev. B 22, 5490 (1980).

62. O. Gunnarsson and P. Johansson, Int. J. Quantum Chem. 10, 307 (1976).

63. A. Zunger, J.P. Perdew and G.L. Oliver, Solid State Commun. 34, 933 (1980).

64. L.M. Bransomb, Atomic and Molecular Processes, Ed. D.R. Bates (Academic, New York, 1962).

Ion	$\Delta E_{min} - E_a$ (eV)	Electron Affinity (eV)
H^-	1.8	0.754
He^-	-	-
Li^-	∿0	0.62
Be^-	∿0	<0
B^-	3.1	0.3 - 0.8
C^-	?	1.3
N^-	4.4	<0
O^-	?	1.5
F^-	9.0	3.5

Table I: Comparison of the depth of the minimum in the energy curve (relative to the LDA atomic energy, Ref. 63) with electron affinities (Ref. 64).

Atom	r_s	$\bar{n}$	$n_o(o)$	E_B^{UDA}	E_B^{corr}	E_B^{exact}
H	1.6	0.05828	0.02617	-4.1		-5.5
	2.0	0.02984	0.01157	-2.3		-3.3
	3.25	0.0070	0.0018	-0.23		0.0
	3.93	0.0039	0.0008	+0.11		0.6
He	1.6	0.05828	0.02617	-8.19	-8.40	-9.87
	2.0	0.02984	0.01157	-4.92	-5.63	-6.04
	3.0	0.00884	0.00243	-1.96	-2.42	-2.26
	4.0	0.00373	0.00073	-0.90	-1.17	-1.03
Li	3.28	0.0068	0.0017	-1.9		-3.7

Table II: Vacancy binding energies (in eV) for H, He and Li in a Z_v=1 vacancy. $\bar{n}$ is the mean host density and $n_o(0)$ is the host density at the centre of the vacancy. E_B^{UDA} is the binding energy based on the UDA; E_B^{Corr} includes electrostatic (46) and gradient (47) corrections; E_B^{exact} is the exact KS value.

Site	ΔE_{UDA}	ΔE_{es}	ΔE_{grad}	ΔE
Octahedral	-0.14	-2.81	1.15	-1.80
Tetrahedral	0.08	-0.92	0.57	+0.45
Saddle-point	1.12	-2.34	0.42	-0.80

Table III: Energy (in eV) of H in Aℓ. The saddle-point is midway between two octahedral sites. The heat of solution is obtained by adding (½ x the dissociation energy = 2.40 eV, Ref.62) to the last column.

Atom	HF	CI	DFT				Expt.
			α_o^{LDA}	α^{LDA}	α_o^{SIC}	α^{SIC}	
H^-		206.4	-	-	90.5	74.4	
He	1.322	1.383	1.772	1.630	1.44	1.29	1.385
Ne	2.36	2.676	3.415	2.997	2.99	2.45	2.666
Ar	10.59	11.10	17.65	11.80	17.83	11.2	11.09
Kr			27.24	17.72	27.92	17.2	16.75
Xe			46.91	28.27	49.13	27.7	27.32
Li^+	0.189	0.192	0.228	0.214			0.192
Na^+	0.945		1.220	1.071			1.07
K^+	5.32		8.812	5.698			5.74±0.14
Rb^+			14.83	9.429			9.57±0.14
Cs^+			27.75	16.32			16.3±0.2
Be	45.5	37.84	78.54	42.92	84.6	45.0	
Mg	81.24	71.3	118.3	69.79	124.6	73.8	
Ca	185.5	153.9	267.4	145.2	286.8	156	169±17
Sr			339.0	184.0	368	198	186±20
Ba			511.7	264.0	561	287	268±22

Table IV: Dipole polarizabilities of atoms and ions in units of a_o^3. LDA are the results based on the local density approximation (Ref. 14), while SIC include self-interaction corrections (Ref. 18). References to the experimental values and HF and CI results are contained in Ref. 11.

Ion	HF		DFT			
	α_q	γ_∞	α_q^o	α_q	γ_∞^o	γ_∞
Li^+	.112	.262	.138	.141	.264	.262
Na^+	1.52	-5.33	1.87	1.88	-5.46	-5.33
K^+	17.4	-19.86	18.4	17.9	-20.2	-19.9
Rb^+	36.1	-53.01	40.1	38.5	-54.3	-53.0
Cs^+	98.7	-102.5	102	93.6	-106	-102
Be^{2+}	.0152	.188	.0177	.0179	.189	.188
Mg^{2+}	.520	-3.44	.592	.591	-3.51	-3.44
Ca^{2+}	6.99	-14.3	7.78	7.46	-14.6	-14.3
Sr^{2+}	18.0	-39.3	19.6	18.4	-41.6	-39.3
Ba^{2+}	52.2	-78.8	57.2	50.5	-83.7	-78.8
Cu^+	20.0	-21.54	65.6	59.7	-18.9	-7.78
Ag^+	27.7	-43.14	64.6	59.9	-37.7	-25.6
Au^+			91.0	82.0	-73.6	-49.5

Table V: Quadrupole polarizabilities (α_q) in units of a_o^5 and Sternheimer antishielding factors (γ_∞). The HF results are taken from Ref. 53 and the DFT results are from Ref. 49.

Metal (r_s)	$\chi_s(10^{-6}cgs)$	K_s(core)	K_s(valence)	K_s	K_s(expt)
Li (3.28)	1.35	-0.0074	0.0348	0.0274	0.026
Na (3.96)	1.25	-0.006	0.137	0.131	0.116
K (4.86)	1.20	-0.017	0.271	0.254	0.265
Rb (5.2)	1.20	-0.030	0.541	0.511	0.662
Cs (5.64)	1.21	-0.097	0.891	0.794	1.44
Be (1.88)	1.84	-0.0085	0.0436	0.0351	0.0
Mg (2.65)	1.48	-0.006	0.143	0.137	0.110
Ca (3.27)	1.35	-0.053	0.306	0.253	-
Sr (3.56)	1.29	-0.115	0.628	0.513	-
Aℓ (2.06)	1.73	-0.008	0.152	0.144	0.164

Table VI: Knight shifts in %. K_s (core) and K_s (valence) are the core and valence contributions, respectively, to the total Knight shifts, K_s. χ_s is the exchange-enhanced spin susceptibility of a uniform electron gas as given by Gunnarsson and Lundqvist (Ref. 14).

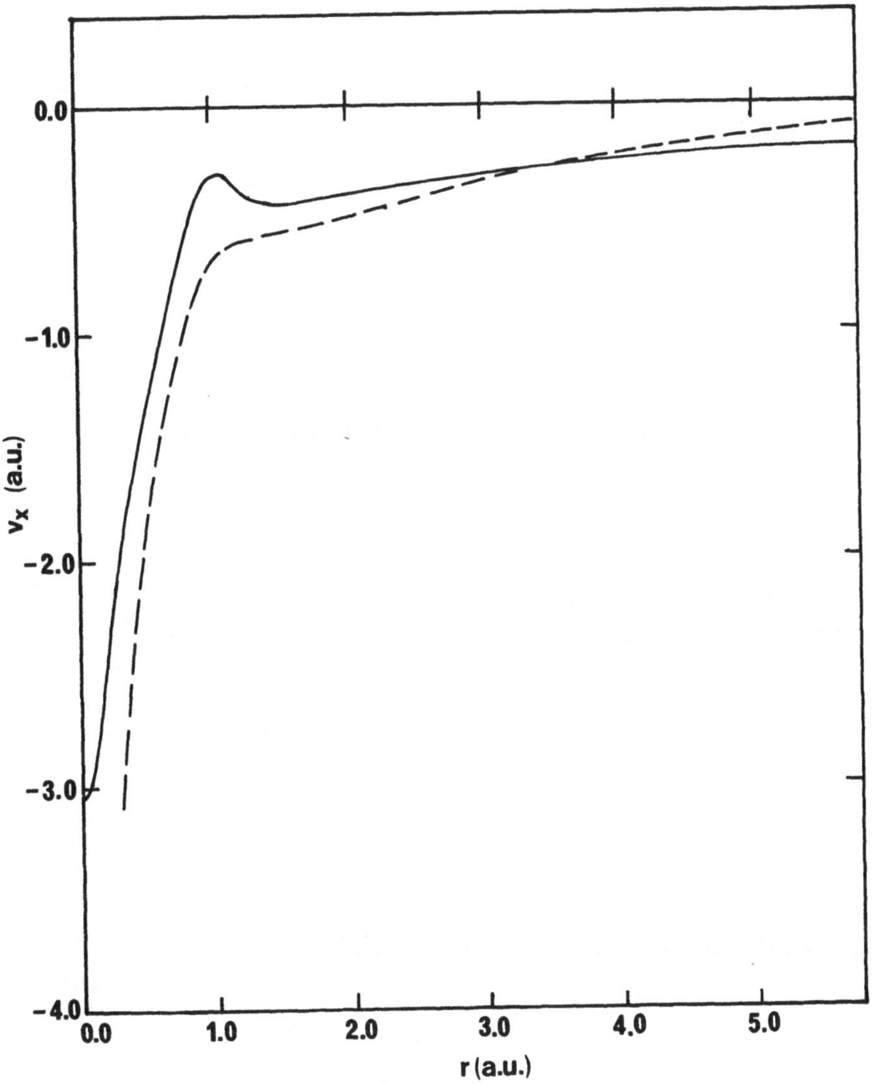

Fig. 1: Comparison of the optimal HF potential (solid line) with the local exchange potential (dashed line) for the atomic density of the optimal potential. Kohn-Sham exchange is used.

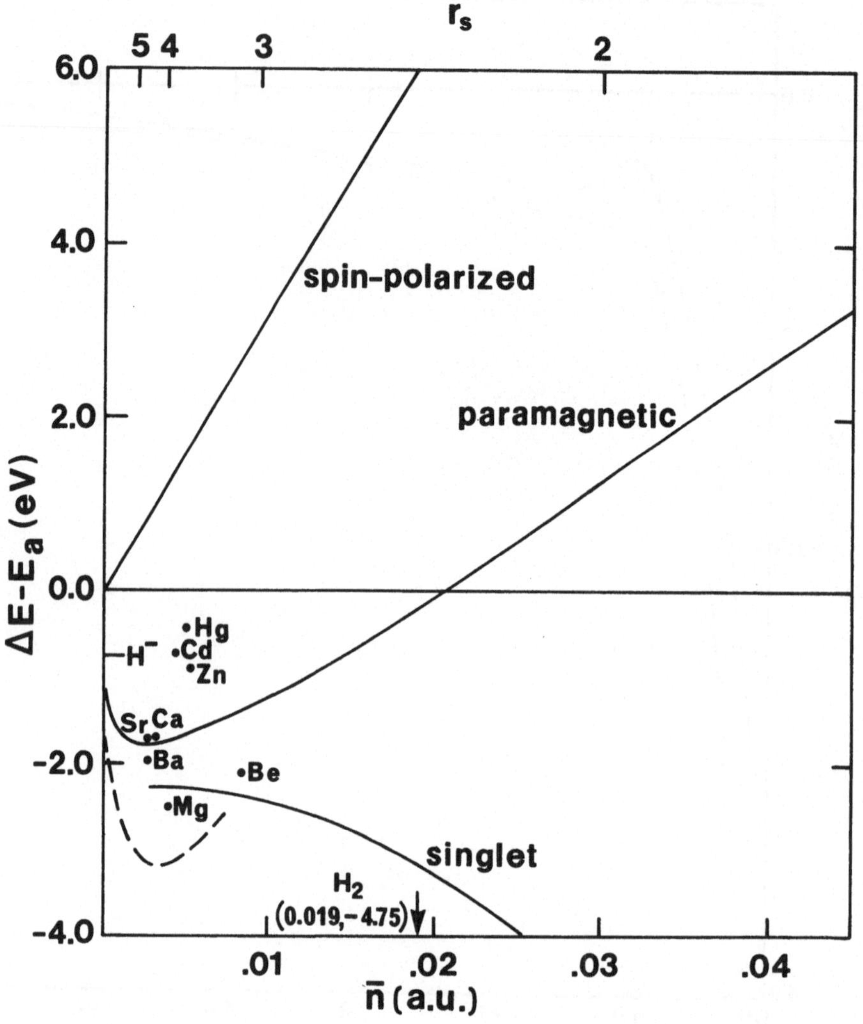

Fig. 2: Energy curves for H in a paramagnetic, spin-polarized and partially polarized (singlet) electron gas. The points represent the energy density coordinates of the various metallic hydrides. The dashed line is a Wigner solid model of the energy minimum.

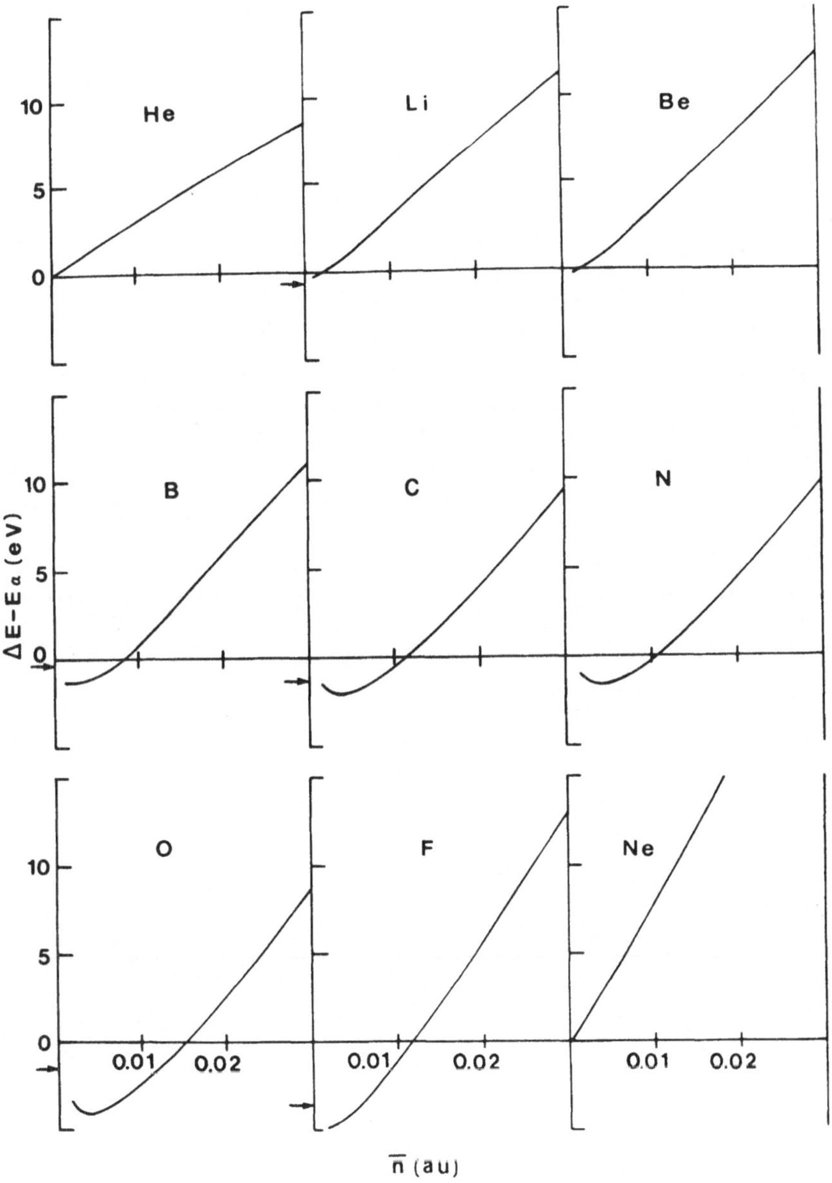

Fig. 3: Quasiatom energy curves for the atoms He through to Ne. The energies are given with respect to the calculated free atom energies. The arrows indicate the energies of the singly charged negative ions.

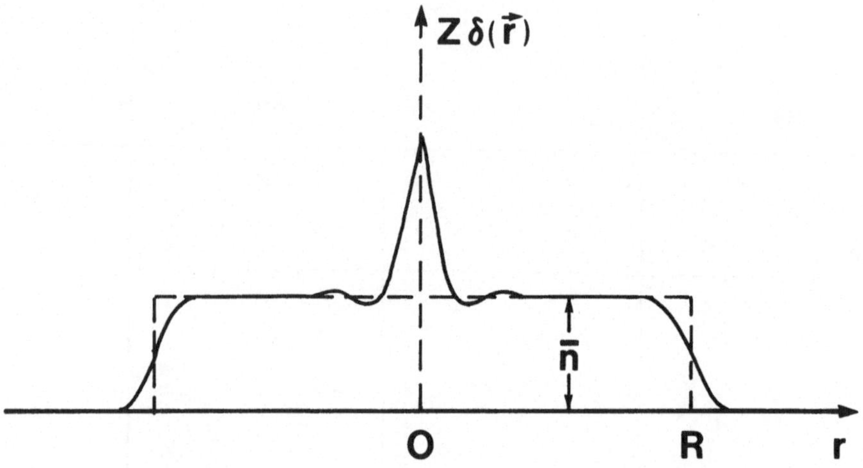

Fig. 4: Schematic of a point charge at the centre of a spherical jellium sample. The dashed lines represent positive charges and the solid line the electronic density.

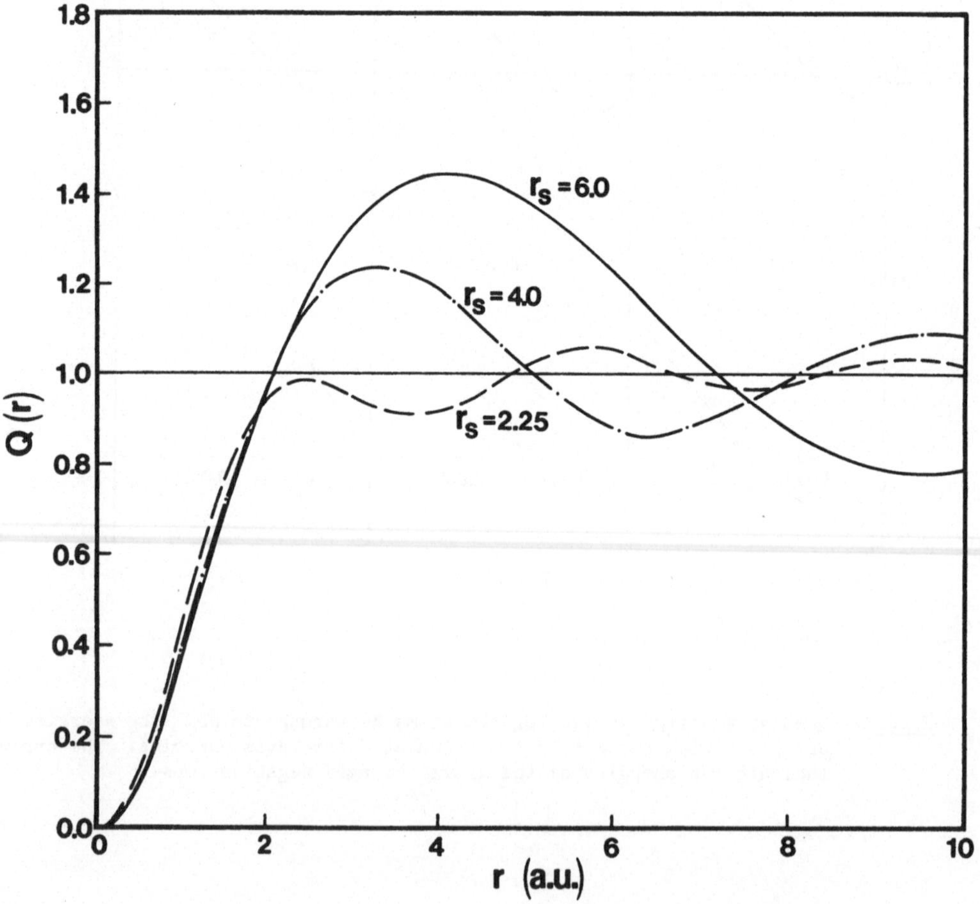

Fig. 5: Integrated charge, Q(r), for H vs. distance for three gas densities.

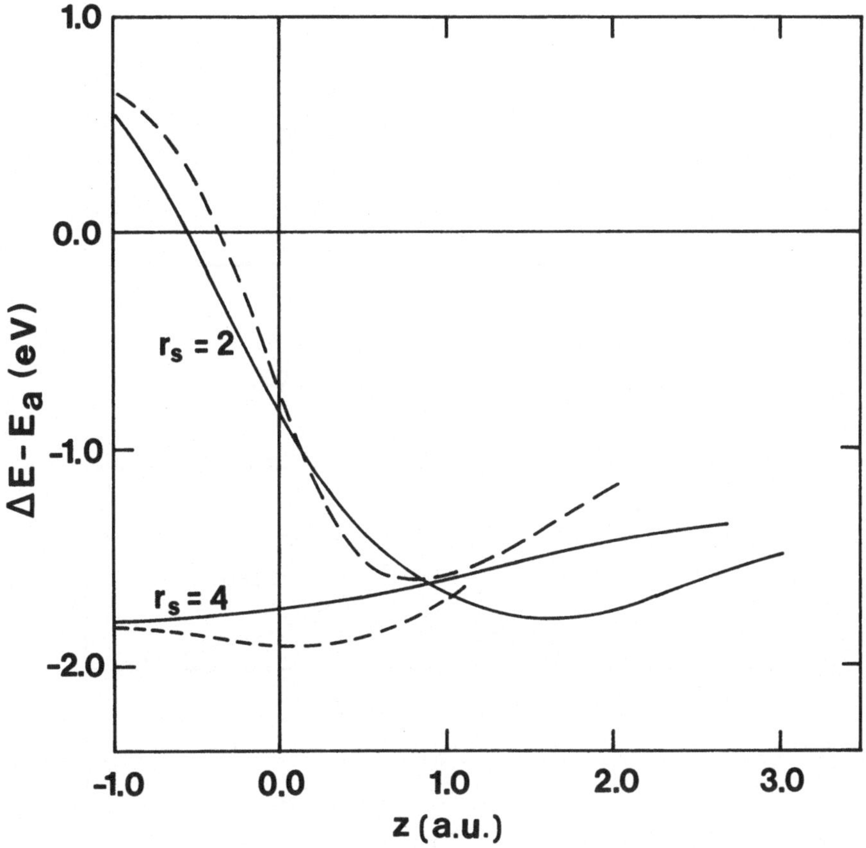

Fig. 6: Chemisorption energy of H on r_s=2 and r_s=4 jellium substrates. The solid line is the UDA estimate and the dashed line is taken from Ref. 29.

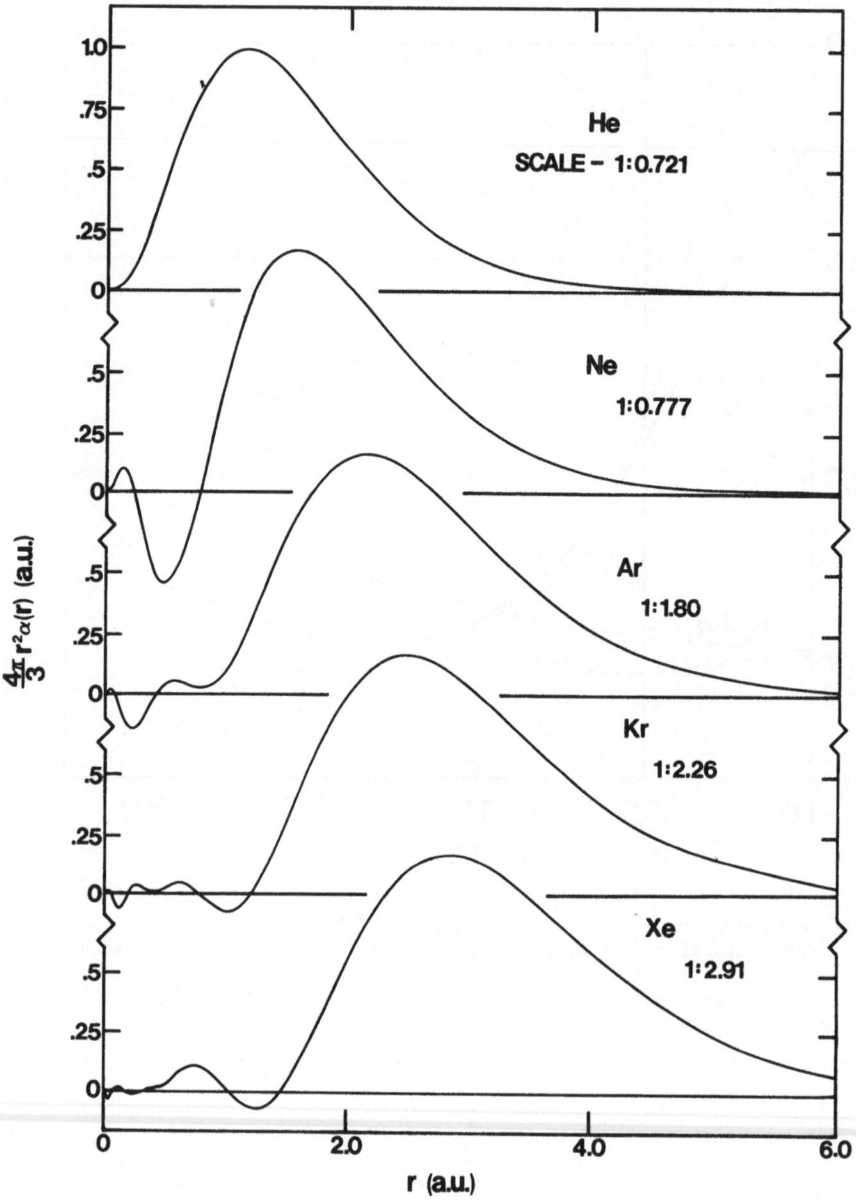

Fig. 7: Induced dipole radial densities for the rare gases.

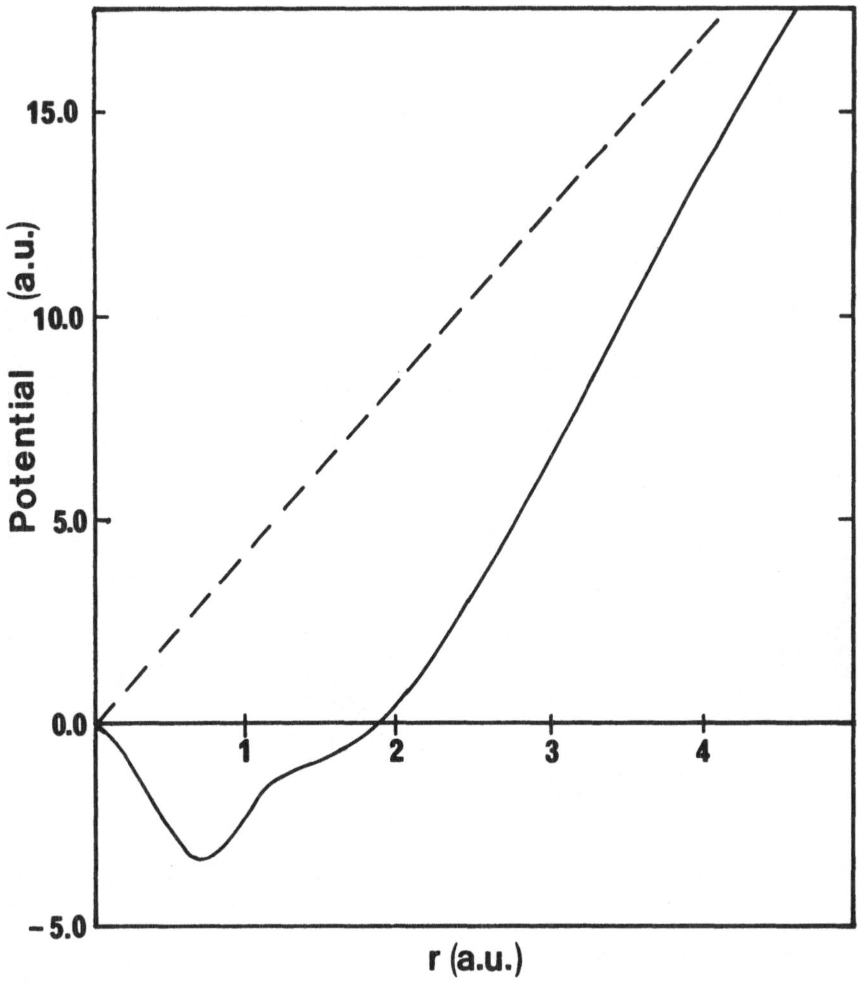

Fig. 8: Comparison of the applied (dashed line) and the total (solid line) perturbing potential for the case of Be.

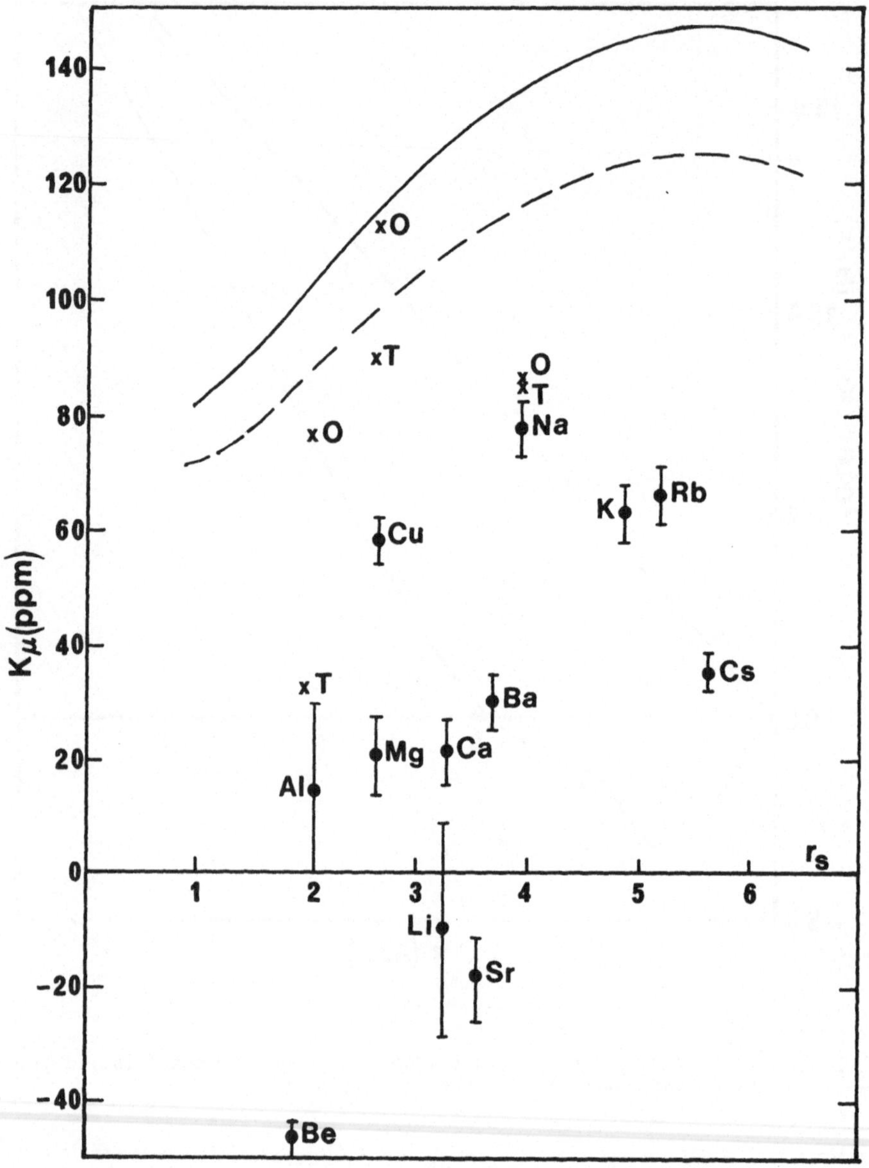

Fig. 9: Comparison of calculated (solid line) Knight shifts with experiment (Ref. 58). The dashed line includes diamagnetic corrections (Ref. 61).

ELECTRON GAS MODELS AND DENSITY FUNCTIONAL THEORY

José Luis Gázquez
D.E.Pg., Facultad de Química
Universidad Nacional Autónoma de México
04510 México, D.F.

CONTENTS:

ABSTRACT.

A functional for representing the kinetic and exchange energies of the ground state of an N electron atom or ion in terms of its electron density is derived by: (1) considering explicitly the variation of the charge density in an atom, and (2) approximating the pair correlation function for parallel spin electrons by that of a uniform free-electron gas, but including the corrections to the momentum at the Fermi level and the appropriate boundary conditions that result from taking into account that the number of electrons in an atom is finite. The first consideration leads in a natural way to the full Weizsacker correction in the kinetic energy expression, while the second consideration yields the Thomas-Fermi term (kinetic) and the Dirac plus the Fermi-Amaldi terms (exchange) multiplied by different functions of N. The quality of such functional is analyzed using Hartree-Fock densities.

I. INTRODUCTION.

The electronic density is one of the most fundamental quantities in quantum chemistry. Many properties of atoms and molecules can be interpreted in terms of the charge distribution and thus, it would be highly desirable to use it as a basic variable, without having to calculate wave functions previously.

The first attempt to relate the energy of a system with its

electronic charge, was made by Thomas[1] and Fermi[2] (TF) in 1927. The idea consisted in approximating the behavior of an atom by that of the uniform electron gas. In this electron gas approximation one describes a system of electrons only by their probability distribution in 3-dimensional space, rather than by a wave function in 3N-dimensional space. One further assumes that, at any point in the distribution, the system behaves locally the same way as a uniform electron gas, with a density equal to that of the system at that point. This way, the whole problem is reduced to that of finding directly the density distribution. In TF theory, the kinetic energy is treated in this manner, and the electron-electron interaction energy consists only of the coulombic repulsion, the exchange and correlation energies are neglected[3,4]. Thus, although the total energy obtained by this method, approaches the exact value[5] in the limit $Z \to \infty$, it is found to be in poor agreement with the Hartree-Fock (HF) results for systems containing a finite number of electrons, atoms cannot bind to form molecules[6,7], and it leads to an atomic charge density which is divergent at the nucleus (instead of being finite), decays as r^{-6} (instead of doing it exponentially) and does not show shell structure[3,4].

In 1930, Dirac[8] included the exchange energy of a uniform electron gas, but this showed only a minor improvement over the TF method. In those days, this electron gas approximation was mainly directed to obtain the electronic density distribution. With the later success of more accurate methods, such as the HF approach, the approximation lost this primary role.

However, there were many efforts to incorporate the inhomogeneity effects of real systems into the electron gas model. In 1935, Weizsacker[9] included a correction to the kinetic energy in terms of the gradient of the charge density. Later, the inhomogeneous electron gas problem, was set up in terms of the quantum corrections that arise because the density matrix is a function of the noncommuting momentum and potential energy operators[10]. Such corrections involved powers of the gradient of the charge density. In particular, the second order correction to the kinetic energy, that results from this approach is 1/9 of the Weizsacker term.

In 1964, Hohenberg and Kohn[11] provided a formal theoretical justification for the energy density functional approach, by demonstrating that the total energy of a many-electron system in its ground state, is a unique functional of its electronic charge density. They were also able to demonstrate the existence of a variational principle for the energy functional.

$$E[\rho] = \int v(\underline{r})\rho(\underline{r})d\underline{r} + F[\rho] \qquad \text{(I-1)}$$

where $\rho(\underline{r})$ represents the electronic charge density, $v(\underline{r})$ is the external potential particular to a given system, and $F[\rho]$ is a universal functional. It was shown that $E[\rho]$ attains its minimum value for the correct $\rho(\underline{r})$ if the trial densities are constrained to yield the correct number of particles N. The universal functional $F[\rho]$ may be expressed as

$$F[\rho] = T[\rho] + V_{ee}[\rho] \qquad \text{(I-2)}$$

$T[\rho]$ and $V_{ee}[\rho]$ being the kinetic and electron-electron interaction energy functionals. Since the classical Coulomb contribution to V_{ee} in terms of ρ is given by

$$J[\rho] = \frac{1}{2}\iint \frac{\rho(\underline{r})\rho(\underline{r}')}{|\underline{r} - \underline{r}'|} d\underline{r}d\underline{r}' , \qquad \text{(I-3)}$$

one can write

$$V_{ee}[\rho] = J[\rho] + K[\rho] \qquad \text{(I-4)}$$

where $K[\rho]$ is the exchange-correlation energy density functional. Unfortunately, except for the trivial case of one-electron systems, the explicit forms of $T[\rho]$ and $K[\rho]$ are not as yet known.

Hohenberg and Kohn considered the gradient expansion of $T[\rho] + K[\rho]$, discussed the conditions under which such an expansion could be expected to converge and gave, from elementary considerations of invariance, the expression of the energy density functional up to fourth order. Besides, using the random phase expression for the electronic polarizability, they found the coefficient of the second order correction to the kinetic energy to be, also, 1/9 of the Weizsacker term.

Since the work of Hohenberg and Kohn, much effort has been devoted towards testing the accuracy of the gradient expansion for calculating the total energy of atoms. In particular, it has been shown[12-15] that when a HF or even better charge density is used, the kinetic energy may be computed quite accurately from the truncated expansion

$$T[\rho] = T_o[\rho] + T_2[\rho] + T_4[\rho] \qquad \text{(I-5)}$$

where the term T_o is the kinetic energy of a free-electron gas,

$$T_o[\rho] = T_{TF}[\rho] = \frac{3}{10}(3\pi^2)^{2/3}\int \rho^{5/3}(\underline{r})d\underline{r}, \tag{I-6}$$

T_2 is 1/9 of the original Weizsacker correction

$$T_2[\rho] = \frac{1}{9}T_w[\rho] = \frac{1}{9}\left\{\frac{1}{8}\int \frac{|\nabla\rho(\underline{r})|^2}{\rho(\underline{r})}d\underline{r}\right\} \tag{I-7}$$

and T_4 is the fourth order correction as given by Hodges[16], namely,

$$T_4[\rho] = \frac{1}{540(3\pi^2)^{2/3}}\int \rho^{1/3}\left[\left(\frac{\nabla^2\rho}{\rho}\right)^2 - \frac{9}{8}\left(\frac{\nabla^2\rho}{\rho}\right)\left(\frac{\nabla\rho}{\rho}\right)^2 + \frac{1}{3}\left(\frac{\nabla\rho}{\rho}\right)^4\right]d\underline{r}. \tag{I-8}$$

Higher-order terms in the expansion of $T[\rho]$ diverge because of the long-range behavior of the charge density[17,18].

Now, within the Hohenberg-Kohn framework, the ultimate goal is to obtain an energy functional whose variation with respect to ρ will yield an Euler equation, which will determine E and ρ for a particular $v(\underline{r})$. In this sense, the truncated expansion for the kinetic energy, gives unsatisfactory results because the functional derivative of the fourth-order correction $\delta T_4/\delta\rho$, will diverge for r approaching infinity, also because of the long range behavior of the charge density.

A similar situation applies for the exchange energy, which can be computed quite accurately from the truncated expansion[19-21]

$$K[\rho] = K_o[\rho] + K_2[\rho] \tag{I-9}$$

where the term K_o is the exchange energy of a free electron gas

$$K_o[\rho] = K_D[\rho] = -\frac{3}{4}\left(\frac{3}{\pi}\right)^{1/3}\int \rho^{4/3}(\underline{r})d\underline{r} \tag{I-10}$$

and K_2 is given by

$$K_2[\rho] = -\frac{3}{4}\left(\frac{3}{\pi}\right)^{1/3}\beta\int \frac{|\nabla\rho(\underline{r})|^2}{\rho^{4/3}(\underline{r})}d\underline{r} \tag{I-11}$$

β being a constant of the order of 10^{-3} which can be determined in several ways. Higher-order terms in the expansion of $K[\rho]$ and also, $\delta K_2/\delta\rho$ diverge because of the long range behavior of the charge density. Thus, although the functional based on Eqs.(I-5) and (I-9) yields accurate energies when accurate densities are used, the presence of T_4 and K_2 would impart physically unacceptable properties to the Euler equation. This problem may be avoided by modeling ρ itself, thereby imposing additional constraints on the variation. Such an alternative has been tested by demanding the density to be a piecewise exponentially decreasing function of the distance from the nucleus[22] or by using a density constructed from a minimal basis set[23]. In both cases, the results showed a marked improvement in the estimates of the total energy obtained by the TF method and its modifications. However, the density thus obtained and consequently the kinetic energy and exchange energy densities were still seriously in error. Besides, this approach leads to a reduction in the generality and applicability of the resulting Euler equation.

Thus, although during a long time, the starting point to relate the energy of a system with its electronic charge has been the electron gas model and the inhomogeneous effects have been treated basically as corrections, it was recently recognized[17], that one should seek for a functional containing only physically acceptable terms and that will yield upon variation a well-behaved Euler equation. In this sense, the homogeneous electron gas terms should be incorporated as corrections rather than as a first-order approximation[24].

Such approach emphasizes the essential role played by the full Weizsacker term in contrast with the gradient expansion, because it is, by itself, the exact kinetic energy functional for one-electron atoms and two-electron HF atoms, it will deliver an Euler equation which will produce the correct long-range behavior, and it has been shown to be a natural component of the exact $T[\rho]$ [25,26].

The plan of this paper is as follows: in Sec. II we review some properties of the first and second order reduced density matrices, and from these, we derive alternative expressions for $T[\rho]$ and $K[\rho]$ at the HF level, in terms of the full Weizsacker term (in the case of $T[\rho]$) and the pair correlation function between parallel spin electrons. We then proceed, in Sec. III, to approximate the pair correlation function by the expression corresponding to an homogeneous electron gas, but modified through the corrections to the momentum at the Fermi level, arising because of taking into account that the number of electrons in an atom or a molecule is finite[27]. This leads to an understanding of the explicit dependence in the number of electrons in the system of the kinetic and exchange energy density functionals of the electron gas type.

II. DENSITY MATRICES AND ENERGY FUNCTIONALS.

It is well known that the total non-relativistic energy of an atom can be expressed as

$$E = -\frac{1}{2}\int \nabla^2\gamma(1,1')\Big|_{1'\to 1} d\tau_1 - Z\int \frac{\gamma(1,1)}{r_1}\, d\tau_1 + \frac{1}{2}\iint \frac{\Gamma(1,2|1,2)}{r_{12}}\, d\tau_1 d\tau_2 \qquad \text{(II-1)}$$

where $\gamma(1,1')$ is the first order reduced density matrix

$$\gamma(1,1') = N \int \Psi^*(1,2,3,\ldots,N)\,\Psi(1',2,3,\ldots,N)\,d\tau_2\ldots d\tau_N dS_1\ldots dS_N \qquad \text{(II-2)}$$

Here, $d\tau_i$ stands for the volume element of the ith electron and dS_i for the spin coordinate. And $\Gamma(1,2|1',2')$ is the second order reduced density matrix

$$\Gamma(1,2|1'2') = N(N-1) \int \Psi^*(1,2,3,\ldots,N)\,\Psi(1',2',3,\ldots,N)\,d\tau_3\ldots d\tau_N dS_1\ldots dS_N \qquad \text{(II-3)}$$

The diagonal elements of the one- and two-particle density matrices are

$$\rho(1) = \gamma(1,1) \qquad \text{(II-4)}$$

and

$$\Gamma(1,2) = \Gamma(1,2|1,2) \qquad \text{(II-5)}$$

$\rho(1)$ gives the charge density at point 1 and is the probability of finding any of the N electrons at this point. Eq. (II-4) through Eq. (II-2) defines this as the number of electrons times the probability of finding a specific electron at point 1. $\Gamma(1,2)$ is the pair density and gives the probability of finding any of the N particles at point 1 and simultaneously another at point 2. According to Eq. (II-5) through Eq. (II-3) this is equal to the number of pairs times the probability of finding simultaneously a specific pair at points 1 and 2.

At the HF level, the first order reduced density matrix is given by

$$\gamma_{HF}(1,1') = \sum_{k=1}^{N} \phi_k^*(1)\phi_k(1') \qquad \text{(II-6)}$$

and the second order reduced density matrix is given by

$$\Gamma_{HF}(1,2|1',2') = \gamma_{HF}(1,1')\gamma_{HF}(2,2') - \gamma_{HF}(1,2')\gamma_{HF}(2,1') \qquad \text{(II-7)}$$

ϕ_k is the kth HF orbital. Substituting Eqs. (II-6) and (II-7) into Eq. (II-1) one obtains the HF total energy as

$$E_{HF}[\gamma] = -\frac{1}{2}\int \nabla^2 \gamma_{HF}(1,1')\Big|_{1'\to 1} d\tau_1 - Z\int \frac{\gamma_{HF}(1,1)}{r_1}\,d\tau_1$$

$$+ \frac{1}{2}\iint \frac{\gamma_{HF}(1,1)\gamma_{HF}(2,2)}{r_{12}}\,d\tau_1 d\tau_2 - \frac{1}{2}\iint \frac{\gamma_{HF}(1,2)\gamma_{HF}(2,1)}{r_{12}}\,d\tau_1 d\tau_2 \qquad \text{(II-8)}$$

Where we can identify the third term as the Coulombic energy and the last term as the exchange energy.

It may be shown that $\gamma_{HF}(1,1')$ is idempotent and that its trace is equal to the number of electrons in the system N. Thus, one could make use of the functional (II-8), and carry out a variation subject to the idempotency and normalization constraints to obtain directly the Euler equation for $\gamma_{HF}(1,1')$ without having to calculate the HF orbitals first. However, here we are interested in the functional of the diagonal elements of the first-order reduced density matrix, $\rho(1) = \gamma(1,1)$, and will not discuss this topic any further, the reader is referred to excellent papers on this topic[28-30].

Following McWeeny[31], for a system of definite spin, $\rho(1)$ and $\Gamma(1,2)$ may be broken into their spin components,

$$\rho(1) = \rho_{\uparrow}(1) + \rho_{\downarrow}(1) \tag{II-9}$$

and

$$\Gamma(1,2) = \Gamma_{\uparrow\uparrow}(1,2) + \Gamma_{\uparrow\downarrow}(1,2) + \Gamma_{\downarrow\uparrow}(1,2) + \Gamma_{\downarrow\downarrow}(1,2) \tag{II-10}$$

where, for instance, $\Gamma_{\uparrow\downarrow}(1,2)$ gives the probability of finding an electron with spin-up at point 1 and another with spin down simultaneously at point 2. For correlated electronic motion, the pair functions may be written in terms of the charge density and the correlation factors between parallel spin electrons $f_{\uparrow\uparrow}(1,2)$ and antiparallel spin electrons $f_{\uparrow\downarrow}(1,2)$ as

$$\Gamma_{\uparrow\uparrow}(1,2) = \rho_{\uparrow}(1)\rho_{\uparrow}(2)\ (1 + f_{\uparrow\uparrow}(1,2)) \tag{II-11}$$

and

$$\Gamma_{\uparrow\downarrow}(1,2) = \rho_{\uparrow}(1)\rho_{\downarrow}(2)\ (1 + f_{\uparrow\downarrow}(1,2)) \tag{II-12}$$

The correlation between electrons of unlike spins is not considered in HF theory, where it is assumed that $f_{\uparrow\downarrow}(1,2) = f_{\downarrow\uparrow}(1,2) = 0$. Thus,

$$\Gamma(1,2) = \rho(1)\rho(2) + \rho_{\uparrow}(1)\rho_{\uparrow}(2)f_{\uparrow\uparrow}(1,2) + \rho_{\downarrow}(1)\rho_{\downarrow}(2)f_{\downarrow\downarrow}(1,2) \tag{II-13}$$

and comparing with Eq.(II-7) we can see that the first order reduced density matrix, $\gamma_{HF}(1,1') \equiv \gamma(1,2)$, may be expressed as (for spin-up electrons)

$$\gamma_{\uparrow\uparrow}(1,2) = \{ - \rho_{\uparrow}(1)\rho_{\uparrow}(2)f_{\uparrow\uparrow}(1,2)\}^{1/2} \tag{II-14}$$

with a similar expression for spin-down electrons. This particular form of the first-order reduced density matrix plays a central role in the present discussion, and thus it is important to look at some of the properties of the product $\rho_{\uparrow}(2)f_{\uparrow\uparrow}(1,2)$.

Because of Pauli exclusion principle, $\Gamma_{\uparrow\uparrow}(1,1) = 0$, and therefore, from Eq.(II-11), we can see that

$$\rho_{\uparrow}(1)f_{\uparrow\uparrow}(1,1) = -\rho_{\uparrow}(1) \tag{II-15}$$

or simply

$$f_{\uparrow\uparrow}(1,1) = -1 \tag{II-16}$$

Since

$$\int \Gamma_{\uparrow\uparrow}(1,2)d\tau_2 = (N_{\uparrow} - 1)\rho_{\uparrow}(1) \tag{II-17}$$

we have that

$$\int \rho_{\uparrow}(2)f_{\uparrow\uparrow}(1,2)d\tau_2 = -1 \tag{II-18}$$

On the other hand, for large r_{12} one may assume that the electrons move independently and therefore $\Gamma_{\uparrow\uparrow}(1,2) \approx \Gamma_{\uparrow\uparrow}^{ind}(1,2)$, where

$$\Gamma_{\uparrow\uparrow}^{ind}(1,2) = \rho_{\uparrow}(1)\rho_{\uparrow}(2) - \rho_{\uparrow}(1)\rho_{\uparrow}(2)/N_{\uparrow} \tag{II-19}$$

to preserve the correct normalization in finite systems. Thus

$$f_{\uparrow\uparrow}(1,2) \to -\frac{1}{N_{\uparrow}} \quad \text{as} \quad r_{12} \to \infty . \tag{II-20}$$

This long-range behavior of $f_{\uparrow\uparrow}(1,2)$ will become very important for systems with a small number of electrons.

Now we proceed to study the expression (II-14) with the properties just found in connection with the kinetic and exchange energy density functionals.

A. Exchange Energy Functional.

According to Eq.(II-14) and the last term of Eq.(II-8), the exchange energy functional may be expressed as

$$K_{\uparrow}[\rho] = \frac{1}{2}\iint \frac{\rho_{\uparrow}(1)\rho_{\uparrow}(2)f_{\uparrow\uparrow}(1,2)}{r_{12}} d\tau_1 d\tau_2 \tag{II-21}$$

or alternatively as

$$K_{\uparrow}[\rho] = \frac{1}{2}\int \rho_{\uparrow}(1)U_{\uparrow}(1)d\tau_1 \tag{II-22}$$

where

$$U_{\uparrow}(1) = \int \frac{\rho_{\uparrow}(2)f_{\uparrow\uparrow}(1,2)}{r_{12}} d\tau_2 \tag{II-23}$$

is an exchange potential set up by an exchange charge

$$\rho_{\uparrow}^{ex}(1,2) = \rho_{\uparrow}(2)f_{\uparrow\uparrow}(1,2) \quad \text{(II-24)}$$

This equation together with the properties defined by Eqs.(II-16) and (II-17) constitutes what is usually called the Fermi-hole, which can be interpreted as a region carried by each electron, forbidden for other electrons of the same spin. To obtain the exchange energy functional it would be necessary to know $f_{\uparrow\uparrow}(1,2)$ in terms of ρ. However, even without specifying the pair correlation function, one may analyze some aspects as follows.

In view of the long-range behavior of $f_{\uparrow\uparrow}(1,2)$ as given by Eq. (II-20), one could consider that for like-spin electrons, $(f_{\uparrow\uparrow} + 1/N_{\uparrow})$ is really the correlation factor. Such consideration, leads directly to the correct long-range behavior of the exchange potential variationally determined from

$$U_{\uparrow}(1) = U_{\uparrow}^{a}(1) + U_{\uparrow}^{b}(1) = \int \frac{\rho_{\uparrow}(2)f_{\uparrow\uparrow}(1,2)}{r_{12}} d\tau_2 - \frac{1}{N_{\uparrow}} \int \frac{\rho_{\uparrow}(2)}{r_{12}} d\tau_2 \quad \text{(II-25)}$$

since

$$V_{\uparrow}^{b}(1) = \frac{\delta(\frac{1}{2}\rho_{\uparrow}U_{\uparrow}^{b})}{\delta\rho_{\uparrow}} = -\frac{1}{N_{\uparrow}} \int \frac{\rho_{\uparrow}(2)}{r_{12}} d\tau_2 \quad \text{(II-26)}$$

and when $r_1 \to \infty$ we have from this equation that

$$V_{\uparrow}^{b}(1) \to -\frac{1}{r_1} \quad \text{(II-27)}$$

Although the long-range behavior of the potential $V_{\uparrow}^{a}(1)$ has not been established, we can look, for example, to its behavior in X_α theory[32], where $f_{\uparrow\uparrow}(1,2)$ is approximated by that of a free electron gas leading to $U_{\uparrow}^{a}(1) \sim \rho_{\uparrow}^{1/3}(1)$ and therefore, $V_{\uparrow}^{a}(1) \sim \rho_{\uparrow}^{1/3}(1)$ which indicates that it will go to zero exponentially when $r_1 \to \infty$ because of the long-range behavior of the charge density. Thus, we will have that $(V_{\uparrow}^{a}(1) + V_{\uparrow}^{b}(1)) \to -1/r_1$ as $r_1 \to \infty$, which is in agreement with the HF exchange potential.

Now, if one separates the total HF exchange potential into self-exchange and interelectronic exchange components, one finds that while the self-exchange potential goes as $-1/r_1$ when $r_1 \to \infty$, the interelectronic exchange potential goes to zero[33]. Thus, we can see that $U_{\uparrow}^{b}(1)$ is closely related with the self-exchange that must cancel with the self-interaction incorrectly included in the Coulombic term. This is most clearly seen if one assumes that $\rho_{\uparrow}(1)/N_{\uparrow} \approx \phi_i^*(1)\phi_i(1)$ because then, the potential given by Eq.(II-26) is exactly equal to the HF self-exchange potential. Such assumption is very poor, however, it enables to see the physical significance of the term that gives rise to the long-range behavior of the pair correlation function between parallel spin electrons.

These results emphasize the importance of the factor $1/N_{\uparrow}$, specially for the light atoms where the self-exchange dominates over the interelectronic exchange, in the helium atom, the whole of the exchange energy is self-exchange energy.

It is interesting to note that the exchange energy arising because of the factor of $1/N_{\uparrow}$ is identical to the one proposed long ago by Fermi and Amaldi[34] to correct for self-interaction in the Thomas-Fermi approximation, since for the spin-restricted case ($\rho_{\uparrow} = \rho_{\downarrow} = \rho/2$), the presence of such term leads to a contribution of the form

$$E_x^b = -\frac{1}{2N} \iint \frac{\rho(1)\rho(2)}{r_{12}} d\tau_1 d\tau_2 \ . \qquad \text{(II-28)}$$

B. Kinetic Energy Functional.

As we have already seen, the kinetic energy density $t[\rho]$ which is related with the kinetic energy functional $T[\rho]$ by $T[\rho] = \int t[\rho]\, d\tau$, may be expressed in terms of the nondiagonal elements of the first-order density matrix as

$$t'[\rho] = -\frac{1}{2} \nabla_{1'}^2 \gamma(1,1')\Big|_{1' \to 1} \qquad \text{(II-29)}$$

or alternatively as

$$t[\rho] = \frac{1}{2} \nabla_1 \nabla_{1'} \gamma(1,1')\Big|_{1' \to 1} \qquad \text{(II-30)}$$

while both densities integrate to the correct total kinetic energy, they differ locally and are related by[35]

$$t'[\rho] = t[\rho] - \frac{1}{4} \nabla^2 \rho(1) \qquad \text{(II-31)}$$

$t[\rho]$ has the advantage over $t'[\rho]$ of being always finite, greater than zero and a smoothly varying function of the coordinates.

In order to analyze the kinetic energy density, we define

$$G(1,1') = \left[-f_{\uparrow\uparrow}(1,1') \right]^{1/2} \qquad \text{(II-32)}$$

so that Eq.(II-14) becomes

$$\gamma_{\uparrow\uparrow}(1,1') = \rho_{\uparrow}^{1/2}(1)\rho_{\uparrow}^{1/2}(1')\, G(1,1') \qquad \text{(II-33)}$$

Substituting this into Eq.(II-30) one obtains that

$$t_{\uparrow}[\rho] = \frac{1}{8} G(1,1) \frac{|\nabla_1 \rho_{\uparrow}(1)|^2}{\rho_{\uparrow}(1)} + \frac{1}{2} \rho_{\uparrow}(1) \nabla_1 \nabla_{1'} G(1,1') \Big|_{1' \to 1}$$

$$+ \frac{1}{2} \rho_{\uparrow}^{1/2}(1) \nabla_1 [\rho(1)]^{1/2} \left[\nabla_1 G(1,1') + \nabla_{1'} G(1,1')\right]_{1' \to 1}$$

(II-34)

Now, since $f_{\uparrow\uparrow}(1,1) = -1$ it follows that

$$G(1,1) = 1 \qquad \text{(II-35)}$$

Besides, one can see that

$$\left[\nabla_1 \gamma(1,1') + \nabla_{1'} \gamma(1,1')\right]_{1' \to 1} = G(1,1) \nabla_1 \rho(1)$$

$$+ \rho(1) \left[\nabla_1 G(1,1') + \nabla_{1'} G(1,1')\right]_{1' \to 1}$$

(II-36)

but, because of Eq.(II-6),

$$\left[\nabla_1 \gamma(1,1') + \nabla_{1'} \gamma(1,1')\right]_{1' \to 1} = \nabla_1 \rho(1) \qquad \text{(II-37)}$$

and therefore,

$$\left[\nabla_1 G(1,1') + \nabla_{1'} G(1,1')\right]_{1' \to 1} = 0 \qquad \text{(II-38)}$$

where we have used Eq.(II-35). Substituting Eqs.(II-35) and (II-38) into Eq.(II-34) one obtains

$$t_{\uparrow}[\rho] = \frac{1}{8} \frac{|\nabla_1 \rho_{\uparrow}(1)|^2}{\rho_{\uparrow}(1)} - \frac{1}{4} \rho_{\uparrow}(1) \nabla_1 \nabla_{1'} f_{\uparrow\uparrow}(1,1') \Big|_{1' \to 1} \qquad \text{(II-39)}$$

as the kinetic energy density at the Hartree-Fock level. This equation was recently found by Ludeña[36] following a more general approach than the one presented here.

Thus, we can interpret the HF kinetic energy as composed of two contributions. The first is the Weizsacker term which arises due to the local variation of the density. If one would have started from a free electron gas, it would have not appeared in the kinetic energy expression even if $\rho = N/V$ is replaced by $\rho(\underline{r})$. Besides, it is important to note that through this formulation we have obtained a coefficient of 1/8 in contrast with the coefficient of 1/72 obtained in the gradient expansion approach. The second term involves the pair correlation function between parallel-spin electrons and consequently its contribution to the kinetic energy comes about because of the antisymmetry requirement on the wave function. Thus, we can see that it is closely related with the exchange energy.

Within the HF framework, Eqs.(II-21) and (II-39) are exact representations of the kinetic and exchange energies. However, to obtain the corresponding density functionals one would still need the exact relation between $f_{\uparrow\uparrow}(1,2)$ and the electronic density, which unfortunately, except for the case of two-electron systems, is not as yet known. Nevertheless, we believe that through the approach presented in this section, it has been possible to get some insight into the energy density functional problem from a point of view different to the gradient expansion.

III. ELECTRON GAS MODELS FOR FINITE SYSTEMS.

In this section we will derive approximate kinetic and exchange energy functionals based on the expression for the pair correlation function of a free electron gas. However, this one will include the corrections to the momentum at the Fermi level that arise from taking into account that the system one is dealing with has a finite number of electrons.

A. The Momentum at the Fermi Level.

In the derivation of the Thomas-Fermi kinetic energy, one starts from the energy levels of a free particle in a cubic box,

$$E(n_1, n_2, n_3) = \frac{\pi^2 \nu^2}{2V^{2/3}} \tag{III-1}$$

where

$$\nu^2 = n_1^2 + n_2^2 + n_3^2 \tag{III-2}$$

Because there is one energy level for each set of integers n_1, n_2 and n_3, the total number of occupied states may be approximated by one-eighth of the volume of a sphere with radius ν_F (value of ν at the Fermi level). That is, one assumes that for a system with an infinite number of electrons, ν may be treated as a continuous variable. Since each state can accommodate two electrons of opposite spin, the total number of electrons will be given by

$$N = 2\left[\frac{1}{8}\left(\frac{4}{3}\pi\nu_F^3\right)\right], \tag{III-3}$$

Substituting the value of ν_F given by this equation in Eq.(III-1), one obtains that

$$E_F = \frac{1}{2}(3\pi^2)^{2/3}\left(\frac{N}{V}\right)^{2/3} \tag{III-4}$$

and therefore

$$p_F = (3\pi^2\rho)^{1/3} \tag{III-5}$$

where $\rho = N/V$ is a constant for a free-electron gas and it is replaced by $\rho(\underline{r})$ for an atom or a molecule. For the spin-polarized case, $N = N_\uparrow + N_\downarrow$, where $N_\uparrow$ is the number of spin-up electrons, $\rho = \rho_\uparrow + \rho_\downarrow$, and

$$p_{F_\uparrow} = (6\pi^2 \rho_\uparrow)^{1/3} \tag{III-6}$$

The procedure just described to derive the number of electrons in terms of ν_F is equivalent to

$$N_\uparrow = \frac{1}{8}\int_0^{\nu_F} 4\pi^2\nu d\nu \tag{III-7}$$

(the factor of 2 in Eq.(III-3) is omitted in the spin-polarized case), because ν is treated as a continuous variable. However, for a system with a finite (and sometimes very small) number of electrons, ν describes a discrete space. Therefore, one should really make a summation taking into account the degeneracy of the energy levels. However, the degeneracy of the energy levels of a free particle in a cubic box has a very irregular pattern. To simplify this, one can assume that the number of electrons is very large (but finite), and to approximate the degeneracy by $4\pi\nu^2$ as in the infinite case, Eq.(III-7). That is, for very large ν the different states with the same energy can be placed on the surface of a sphere with radius ν. Thus, one can propose to replace the integral by a sum of the form

$$N_\uparrow = \frac{1}{8}\sum_{\nu=0}^{\nu_F} 4\pi^2\nu^2 \tag{III-8}$$

and therefore

$$N_\uparrow = \frac{\pi}{2}\frac{\nu_F}{6}(2\nu_F + 1)(\nu_F + 1) \tag{III-9}$$

Now, in order to obtain p_F, one needs to express ν_F in terms of $N_\uparrow$. Since $N_\uparrow$ has been assumed to be very large, one can perform an asymptotic expansion in Eq.(III-9) which leads to the relation

$$\nu_F = \left(\frac{6N_\uparrow}{\pi}\right)^{1/3}\left[1 - \frac{1}{2}\left(\frac{\pi}{6}\right)^{1/3} N_\uparrow^{-1/3} + O\left(N_\uparrow^{-2/3}\right)\right] \tag{III-10}$$

and therefore, the momentum at the Fermi level will be given by

$$p_{F_\uparrow} = (6\pi^2\rho_\uparrow)^{1/3}\left[1 - \frac{1}{2}\left(\frac{\pi}{6}\right)^{1/3} N_\uparrow^{-1/3} + O\left(N_\uparrow^{-2/3}\right)\right] \tag{III-11}$$

where the relation $\rho_{\uparrow} = N_{\uparrow}/V$ has been used, and we have kept only the terms up to $N_{\uparrow}^{-1/3}$ in the asymptotic expansion.

Equation (III-11) contains the corrections due to the fact that the number of electrons in the system is finite. These corrections will modify the expression for the kinetic energy, Eq.(I-6), even though one is dealing with an homogeneous electron gas. However, notice that

$$p_{F_{\uparrow}} \rightarrow (6\pi^2\rho_{\uparrow})^{1/3}$$

when $N_{\uparrow} \rightarrow \infty$, which is equal to Eq.(III-6).

Now, although the present derivation involves the assumption of a very large number of electrons in order to perform the asymptotic expansion, and to solve the degeneracy problem, it does allow to establish the $N_{\uparrow}$ dependence of the momentum at the Fermi level, even for small $N_{\uparrow}$.

B. <u>Electron Gas Model for the Kinetic Energy</u>.

Now we make use of Eqs.(II-39) and (III-11) in order to derive an approximate kinetic energy functional.

First we note that Eq.(II-39) requires knowledge of the pair correlation function between parallel-spin electrons. Following the idea of taking into account that the number of electrons in the system is finite, we express $f_{\uparrow\uparrow}(1,2)$ in the form

$$f_{\uparrow\uparrow}(1,2) = -\left[\left(1 - \frac{1}{N_{\uparrow}}\right) C_{\uparrow\uparrow}(1,2) + \frac{1}{N_{\uparrow}}\right] \quad \text{(III-12)}$$

where $C_{\uparrow\uparrow}(1,2)$ must be such that

$$C_{\uparrow\uparrow}(1,1) = 1 \quad \text{(III-13)}$$

and

$$C_{\uparrow\uparrow}(1,2) \rightarrow 0 \quad \text{as} \quad r_{12} \rightarrow \infty \quad \text{(III-14)}$$

to satisfy the conditions given by Eqs.(II-16) and (II-20). That is, the long range behavior of $f_{\uparrow\uparrow}(1,2)$ is considered explicitly. Since $f_{\uparrow\uparrow}(1,2) \rightarrow C_{\uparrow\uparrow}(1,2)$ as $N_{\uparrow} \rightarrow \infty$, one can approximate $C_{\uparrow\uparrow}(1,2)$ by the pair correlation function for an infinite free-electron gas, therefore

$$C_{\uparrow\uparrow}(1,2) = 9\left(\frac{\sin x - x\cos x}{x^3}\right)^2 \quad \text{(III-15)}$$

where $x = p_{F_{\uparrow}} r_{12}$. This relation satisfies Eqs.(III-13) and (III-14) and one should use it together with Eqs.(III-11) and (III-12) to describe an homogeneous electron gas with a finite number of electrons. There have been other approaches[12,37] to account for the finite number of electrons in an atom that consider, really, a finite size box, and treat the degeneracy with no approximation, however, through this procedure is difficult to see the $N_\uparrow$ dependence of the different contributions to the total energy.

Substituting Eq.(III-12) into Eq.(II-39) and using Eq.(III-15) one finds that

$$t_\uparrow[\rho] = \frac{1}{8}\frac{|\nabla_1\rho_\uparrow(1)|^2}{\rho_\uparrow(1)} + \frac{3}{5}\,p_{F_\uparrow}^2\left(1-\frac{1}{N_\uparrow}\right)\rho_\uparrow(1) \tag{III-16}$$

and replacing $p_{F_\uparrow}$ through Eq.(III-11) one has that

$$t_\uparrow[\rho] = \frac{1}{8}\frac{|\nabla_1\rho_\uparrow(1)|^2}{\rho_\uparrow(1)} + \frac{3}{10}(6\pi^2)^{2/3}\rho_\uparrow^{5/3}\left(1-\frac{1}{N_\uparrow}\right)\left[1-\left(\frac{\pi}{6}\right)^{1/3}N_\uparrow^{-1/3}\right] \tag{III-17}$$

where we have kept only the terms up to $N_\uparrow^{-1/3}$ in the expansion of $p_{F_\uparrow}^2$. It should be mentioned that if we would have started from the free-electron gas, only the second term of Eq.(III-17) would have appeared in the kinetic energy density expression. The first term is a consequence of the local variation of the charge density as already mentioned. However, although apparently we are incorporating since the very beginning the inhomogeneity of the real system, this is not completely true, because in order to derive Eq.(III-15) one must assume that the charge density is constant.

Nevetherless, through the present approach, we have been able to obtain a functional that contains the full Weizsacker correction plus the Thomas-Fermi kinetic energy multiplied by a function of the number of electrons[38,39]. The first term will lead to an adequate Euler equation, while the second will lead to the exact functional when $Z = N \to \infty$. Besides, the presence of the factor $(1 - 1/N_\uparrow)$ makes the functional exact for one-electron atoms and two-electron HF atoms.

The functional given by Eq.(III-17) is almost identical to the one proposed by Acharya, Bartolotti, Sears and Parr[24], the difference can be established by defining

$$\gamma(N,Z) = \frac{T[\rho] - T_W[\rho]}{T_{TF}[\rho]} \tag{III-18}$$

Thus, through the present approach, it has been found that (for the spin restricted case),

$$\gamma(N,Z) = (1 - \frac{2}{N}) \left\{1 - \frac{C_o}{N^{1/3}}\right\} \qquad \text{(III-19)}$$

where C_o = 1.015, while they have proposed that since $T_{TF} + T_W$ overestimates $T[\rho]$, and since $T_W[\rho]$ represents a highly desirable component of the functional, one should subtract from $T_{TF}[\rho]$ the excess kinetic energy. Thus, they found, using HF atomic densities, that the kinetic energy can be fitted remarkably well with

$$\gamma(N,Z) = 1 - C_o/N^{1/3} \qquad \text{(III-20)}$$

where C_o is a constant (for 55 neutral atoms C_o = 1.412, with rms error 0.053). They conclude that the Weizsacker T_W approximately represents the kinetic energy of the electrons in the K shell, and therefore, when it is used together with T_{TF}, one should also include the term $-(C_o/N^{1/3})\,T_{TF}$ to subtract the incorrect statistical estimate of that energy.

One can note that the leading term in Eq.(III-19) goes as $N^{-1/3}$, therefore Eq.(III-20) is a particular case of Eq.(III-19). On the other hand, Schwinger[40] has considered the correct treatment of strongly bound electrons within the framework of the complete Thomas-Fermi theory and has found that the leading correction is of the order of $N^{-1/3}$. This result is in complete agreement with the one found through the present approach.

Tal and Bader[17] have also found that the Weizsacker term by itself gives a very good description of the kinetic energy density close to the nucleus (where the charge density varies rapidly), and that $T_{TF}[\rho]$ is required, in addition to $T_W[\rho]$, in the region where ρ is a slowly varying function of the distance from the nucleus. Thus, based on a partitioning of the density $\rho(\underline{r})$ as

$$\rho(\underline{r}) = \rho_1(\underline{r}) + \rho_2(\underline{r}), \qquad \text{(III-21)}$$

where $\rho_1(\underline{r})$ is the rapidly varying component,

$$\rho_1(\underline{r}) = \rho(0)\, e^{-2Zr} \qquad \text{(III-22)}$$

and where $\rho_2(\underline{r})$ is the slowly varying component, they have proposed the kinetic energy functional

$$T[\rho] = \frac{1}{8}\int \frac{|\nabla\rho_1(\underline{r})|^2}{\rho_1(\underline{r})}\,d\tau + \frac{1}{8}\int \frac{|\nabla\rho_2(\underline{r})|^2}{\rho_2(\underline{r})}\,d\tau + \frac{3}{10}(3\pi^2)^{2/3}\int \rho_2^{5/3}(\underline{r})\,d\tau\ . \qquad \text{(III-23)}$$

This way, the TF kinetic energy is neglected in the region where the charge density varies rapidly (close to the nucleus).

Now, one can see that there is a connection between the two models, Eqs.(III-18) and (III-23). The term neglected in Eq.(III-23) must be approximately equal to the term subtracted from Eq.(III-18) for $\gamma(N,Z)$ given by Eq.(III-20), that is

$$\rho^{5/3}(0) \int e^{-10Zr/3} d\tau \approx \frac{C_o}{N^{1/3}} \int \rho^{5/3}(\underline{r}) d\tau \qquad \text{(III-24)}$$

To test such relation, one can carry out a fitting of the form aZ^b to the term $\int \rho^{5/3} d\tau$ evaluated with HF charge densities. Thus, it is obtained a = 0.161 and b = 2.407 (rms = 0.0046), which combined with $C_o/Z^{1/3}$ (C_o = 1.412) gives for the right-hand side of Eq.(III-24) $0.227\ Z^{2.074}$. On the other hand, Tal and Bader have fitted $\rho(0)$ in the form $0.4798\ Z^{3.1027}$, which combined with the result for $\int e^{-10Zr/3} d\tau$, gives for the left-hand side of Eq.(III-24), $0.200Z^{2.171}$, which is in good agreement with the result obtained for the right-hand side. These results establish that Eq.(III-24) is approximately valid, and therefore one should expect both models to be somewhat similar.

In summary, the present discussions come in support of using the full Weizsacker term for the description of the kinetic energy density and to incorporate the Thomas-Fermi term as a correction, in contrast with the gradient expansion approach.

The quality of Eq.(III-17) to predict the kinetic energy when used with HF densities, may be established through Fig.1, where we have plotted Eqs.(III-18) and (III-19). To calculate T, T_W and T_{TF} in Eq.(III-18) we have used the HF orbitals for the neutral atoms $2 \leq Z \leq 54$ of Clementi and Roetti[41]. It may be seen that the model agrees reasonably well with the exact values. A least squares fit to fix the value of the constant C_o gives remarkably good results[27], indicating that the N-dependence obtained through the present approach is basically correct, although the value of C_o still shows some discrepancy.

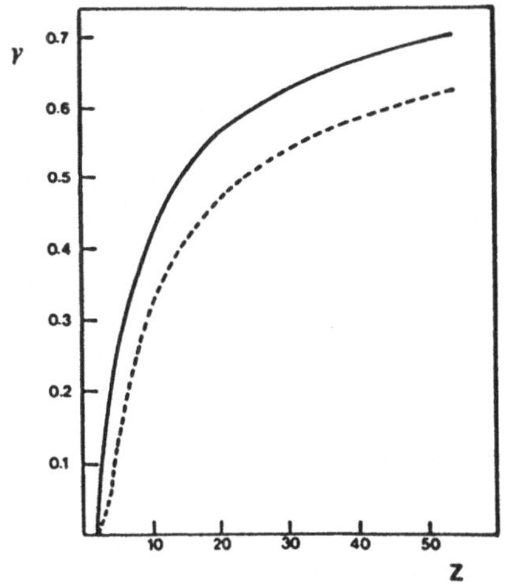

Fig. 1. Continuous line represents $\gamma(N,Z)$ as given by Eq.(III-19) vs. atomic number, 53 neutral atoms. The dashed line is obtained from Eq.(III-18) using HF charge densities.

C. Electron Gas Model for the Exchange Energy.

Now we make use of Eqs.(II-21) and (III-11) in order to derive an approximate exchange energy functional. Since Eq.(II-21) requires knowledge of the pair correlation function between parallel-spin electrons, we make use of Eqs.(III-12) and (III-15) as in the case of the kinetic energy. Thus, we find that the exchange potential, Eq. (II-23), is given by

$$U_{\uparrow}(1) = U_{\uparrow}^{a}(1) + U_{\uparrow}^{b}(1) = -\left(1 - \frac{1}{N_{\uparrow}}\right)\int \frac{\rho_{\uparrow}(2)C_{\uparrow\uparrow}(1,2)}{r_{12}}\,d\tau_2 - \frac{1}{N_{\uparrow}}\int \frac{\rho_{\uparrow}(2)}{r_{12}}\,d\tau_2 \qquad \text{(III.25)}$$

where the second term resembles the self-interaction potential as shown in Sec. II, and the first term represents the interelectronic exchange potential, $C_{\uparrow\uparrow}(1,2)$ is assumed to be given by Eq.(III-15).

To proceed any further is neccessary to know explicitly $\rho_{\uparrow}$ as a function of $\underline{r}_2$, in order to carry out the integrations. This was not the case in the kinetic energy (Sec.II), where we simply wrote formally the derivatives of $\rho_{\uparrow}$ with respect to $\underline{r}_2$ and evaluated them at $\underline{r}_2 = \underline{r}_1$, therefore considering explicitly the local variation of the electronic density. In the present case, one could leave the exchange potential as written in Eq.(III-25), however, in order to develop an approximate functional, we will assume that $\rho_{\uparrow}(2) \approx \rho_{\uparrow}(1)$ as in the free-electron gas case. First, we will consider such approximation only for the first term of Eq.(III-25), $U_{\uparrow}^{a}(1)$, while $U_{\uparrow}^{b}(1)$ will be mantained as given. Thus, in contrast with the kinetic energy density, here we consider only partially the local variation of the charge density through the second term. In the first term, one has also present, implicitly, the approximation through Eq.(III-15), as in the kinetic energy.

This way, we can express the first term of Eq.(III-25) as

$$U^a_{\uparrow}(1) = -\left(1 - \frac{1}{N_{\uparrow}}\right) \rho_{\uparrow}(1) \int \frac{C_{\uparrow\uparrow}(1,2)}{r_{12}} d\tau_2 , \qquad \text{(III-26)}$$

and substituting Eq.(III-15), one obtains after integration that

$$U^a_{\uparrow}(1) = -\left(1 - \frac{1}{N_{\uparrow}}\right) \rho_{\uparrow}(1) \frac{9\pi}{p^2_{F_{\uparrow}}} \qquad \text{(III-27)}$$

However, we should subtract from this expression the statistical estimate of $U^b_{\uparrow}(1)$ to make it approximately self-interaction free and thus corresponding to an interelectronic-exchange potential. This is necessary because in the integration one does not exclude the self-interaction term[42]. Another form of understanding such correction, consists in realizing that the total exchange charge density should integrate to -1, and since the charge $-\rho_{\uparrow}(2)/N_{\uparrow}$ which gives rise to the potential $U^b_{\uparrow}(1)$ already integrates to -1, the charge corresponding to $U^a_{\uparrow}(1)$ should integrate to zero. However, one finds that

$$-\left(1 - \frac{1}{N_{\uparrow}}\right)\rho_{\uparrow}(1) \int C_{\uparrow\uparrow}(1,2) d\tau_2 = -\left(1 - \frac{1}{N_{\uparrow}}\right)\left[1 + \frac{1}{2}\left(\frac{\pi}{6}\right)^{1/3} N_{\uparrow}^{-1/3} + O(N_{\uparrow}^{-2/3})\right] \qquad \text{(III-28)}$$

where we have made use of Eq.(III-11) and have kept only the terms to the order of $N_{\uparrow}^{-1/3}$. Eq.(III-28) is equal to zero only for $N_{\uparrow} = 1$, and goes to -1 as $N_{\uparrow} \to \infty$. This last limit is correct, since for $N_{\uparrow} \to \infty$, the second term of Eq.(III-25) goes to zero and the total exchange potential is given only by the first term which arises because of an exchange charge density that integrates to -1. For any other $N_{\uparrow}$ different than one, Eq.(III-28) is different from zero. Thus, one can propose a correction to the potential given by Eq.(III-27) which consists of the potential set up by a charge $-\rho_{\uparrow}(1)/N_{\uparrow}$, uniformly distributed within a sphere of radius r_o, fixed so that the total charge integrates to -1. This amounts to subtract -1 from Eq.(III-28), so that the value of the integral is approximately given by $-\frac{1}{2}\left(\frac{\pi}{6}\right)^{1/3} N_{\uparrow}^{-1/3}$, which although it is different from zero for finite $N_{\uparrow}$, it has a small value even for small $N_{\uparrow}$. Therefore, we write

$$U^a_{\uparrow}(1) = -\left(1 - \frac{1}{N_{\uparrow}}\right) \rho_{\uparrow}(1) \frac{9\pi}{p^2_{F_{\uparrow}}} + \frac{(36\pi)^{1/3}}{2} \frac{1}{N_{\uparrow}^{1/3}} \rho_{\uparrow}^{1/3}(1) \qquad \text{(III-29)}$$

where the second term corresponds to subtracting the potential mentioned above, and substituting Eq.(III-11) one finally finds that

$$U_{\uparrow}(1) = -\frac{3}{2}\left(\frac{6}{\pi}\right)^{1/3}\rho_{\uparrow}^{1/3}(1)\left[1 - D_o N_{\uparrow}^{-1/3}\right] - \frac{1}{N_{\uparrow}}\int \frac{\rho_{\uparrow}(2)}{r_{12}}\,d\tau_2 \qquad \text{(III-30)}$$

where $D_o = (\pi/6)^{1/3}\,[(4\pi/3)^{1/3} - 1] = 0.4933$ and we have kept only the terms up to $N_{\uparrow}^{-1/3}$ in the expansion of $p_{F_{\uparrow}}^{-2}$.

Thus, we have found an exchange functional that contains the Dirac exchange multiplied by a function of the number of electrons, plus a self-interaction correction, which reduces to the Fermi-Amaldi term for the spin-restricted case and that will lead to the correct long-range behavior of the exchange in the corresponding Euler equation[43].

The quality of Eq.(III-30) to predict the exchange energy when used with HF densities, may be established by defining

$$\delta(N,Z) = \frac{K[\rho] + \frac{1}{N} J[\rho]}{K_D[\rho]}, \qquad \text{(III-31)}$$

where to calculate K, J and K_D one can use the HF orbitals for the neutral atoms $2 \leq Z \leq 54$ of Clementi and Roetti, and comparing it with the one we have obtained, namely,

$$\delta(N,Z) = 1 - D_o' N^{-1/3} \qquad \text{(III-32)}$$

where $D_o' = 0.6215$ for the spin restricted case. We can see from Fig. 2a that there is a fair agreement. However, if D_o' is determined through a least squares fit, the agreement is very good, indicating that the predicted N-dependence is basically correct.

Now, to arrive at Eq.(III-30), we have treated $U_{\uparrow}^a(1)$ and $U_{\uparrow}^b(1)$ on different grounds. Thus, it seems worth to investigate the functional that would result by treating $U_{\uparrow}^b(1)$ through a local approximation of the type $\rho_{\uparrow}(2) \approx \rho_{\uparrow}(1)$, as in the case of $U_{\uparrow}^a(1)$. However, since this would correspond exactly to the term subtracted in Eq.(III-29) to correct for the self-interaction, one would simply have then that the total exchange potential may be expressed as

$$U_{\uparrow}(1) = -\left(1 - \frac{1}{N_{\uparrow}}\right)\rho_{\uparrow}(1)\frac{9\pi}{p_{F_{\uparrow}}^2} = -\frac{3}{2}\left(\frac{6}{\pi}\right)^{1/3}\rho_{\uparrow}^{1/3}(1)\left(1 - \frac{1}{N_{\uparrow}}\right)\left(1 + E_o N_{\uparrow}^{-1/3}\right) \qquad \text{(III-33)}$$

where $E_o = (\pi/6)^{1/3} = 0.806$ and we have kept only the terms up to $N_{\uparrow}^{-1/3}$ in the expansion of $p_{F_{\uparrow}}^{-2}$.

Thus, we have found an exchange functional that contains the Dirac exchange multiplied by a function of the number of electrons only, thus, this functional will not lead to the correct long-range behavior of the exchange potential in the corresponding Euler equation, because it will decay exponentially instead of as $-1/r_1$ when $r_1 \rightarrow \infty$.

Nevertheless, one can test the capacity of such functional to predict the exchange energy, when it is used together with HF densities. As previously, we define

$$\varepsilon(N,Z) = \frac{K[\rho]}{K_D[\rho]} \tag{III-34}$$

where again, K and K_D are calculated with HF orbitals, and the results are compared with

$$\varepsilon(N,Z) \doteq (1 - \frac{2}{N})(1 + E_o' N^{-1/3}) \tag{III-35}$$

where $E_o' = 1.015$. We can see from Fig.2b that there is a good agreement. Again, if E_o' is determined through a least squares fit, the agreement is greatly improved, thereby favoring the predicted N-dependence. In particular, a fit to ε of the form $A + B/N^{1/3}$ yields excellent results as shown by Gadre, Bartolotti and Handy[44]. This indicates that the leading correction to the exchange energy goes as $N^{-1/3}$. Since the leading term in Eq.(III-35) goes as $N^{-1/3}$, the results obtained through the present approach are in agreement with such observations.

It is important to note that Eq.(III-34) is similar to the exchange functional used in the X_α theory, in fact, the parameter α may be expressed in terms of $\varepsilon(N,Z)$ as

$$\alpha = \frac{2}{3}\,\varepsilon(N,Z) \tag{III-36}$$

Thus, through the present approach, it is possible to explain the Z-dependence of the α parameter[45]. This has also been explained through the use of different approximations to the pair correlation function between parallel-spin electrons and the set of conditions that describe the Fermi-hole[46,47] (Eqs.(II-16), (II-18) and (II-20). That is, the use of a model for $f_{\uparrow\uparrow}(1,2)$ is somehow equivalent to the inclusion of the corrections to the momentum at the Fermi level in the free-electron gas $f_{\uparrow\uparrow}(1,2)$ to account for the finite number of electrons.

In summary, the present discussion on the exchange energy functional has shown the possibility of combining the Dirac and Fermi-Amaldi terms, to ensure the correct long-range behavior and to get a better cancellation of the self-interaction, specially at large

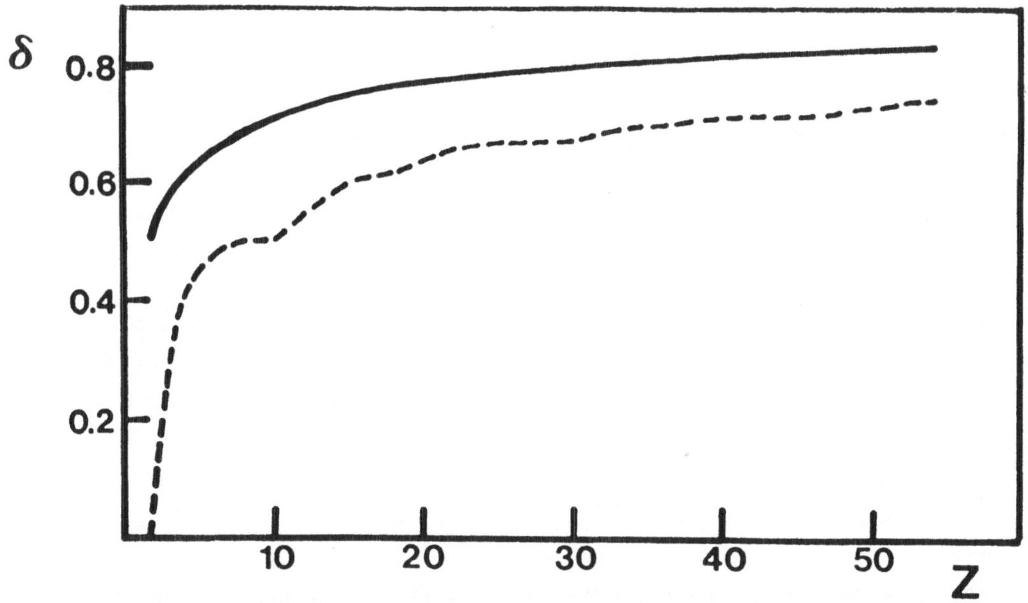

Fig. 2a. Continuous line represents $\delta(N,Z)$ as given by Eq.(III-32) vs. atomic number. The dashed line is obtained from Eq.(III-31) using HF charge densities of 53 neutral atoms.

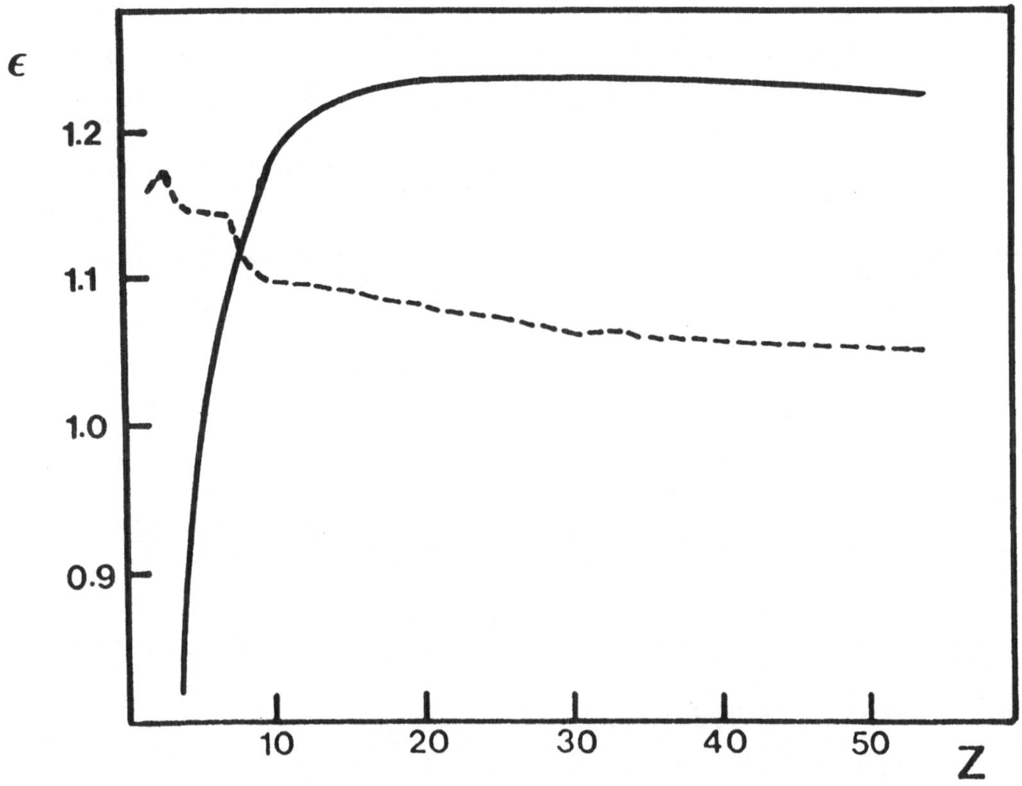

Fig. 2b. Continuous line $\varepsilon(N,Z)$ as given by Eq.(III-35) vs. atomic number. Dashed

distances. On the other hand, such functional could be used within the context of the Kohn-Sham formalism[48], where it could be anticipated to yield results closer to HF than the X_α method, without increasing computing efforts. In fact, we have shown that through the inclusion of the Fermi-Amaldi term into the Kohn-Sham equations, Koopmans theorem is satisfied, so that the eigenvalues approximately represent removal energies[49]. Besides, it is interesting to note that, by a totally different approach, Golden[50] has derived an equation that together with a correction term is a lower bound for HF energies and whose exchange part bears some similarity to Eq.(III-30), the difference being in the function of the number of electrons that multiplies the Dirac term.

D. Euler Equation.

In the previous paragraphs, we have tested the quality of the functionals proposed through the calculation of the kinetic and exchange energy with the HF electronic density. Now we turn to the problem of calculating the charge density through the energy functional. This is a desirable next step. That is, the potential of the energy density functional approach lies in the possibility of determining E and ρ without recourse to any other method, thereby without having to calculate wave functions at all.

The problem consists in carrying out a variation of $E[\rho]$ with respect to $\rho(\underline{r})$ subject to the constraint that the total number of electrons remains fixed, this yields what is called an Euler equation, and then, by solving this equation one obtains the electronic density corresponding to the ground state of the system.

In what follows, we will illustrate how to carry out the variation of the functional

$$E[\rho] = \int \xi[\rho]\, d\tau \tag{III-37}$$

where the total energy density is given by

$$\xi[\rho] = \frac{1}{8}\frac{|\nabla_1\rho(1)|^2}{\rho(1)} + \frac{3}{10}(3\pi^2)^{2/3}\gamma(N,Z)\rho^{5/3}(1) - Z\frac{\rho(1)}{r} + \frac{1}{2}\rho(1)\int\frac{\rho(2)}{r_{12}}d\tau_2 - \frac{1}{2N}\rho(1)\int\frac{\rho(2)}{r_{12}}d\tau_2 - \frac{3}{4}\left(\frac{3}{\pi}\right)^{1/3}\delta(N,Z)\rho^{4/3}(1) \tag{III-38}$$

and the restriction is

$$N[\rho] = \int \rho\, d\tau \tag{III-39}$$

The condition for $E[\rho]$ to be stationary with respect to variation of the charge density is

$$\delta\{E[\rho] - \mu N[\rho]\} = \delta H[\rho] = 0 \qquad \text{(III-40)}$$

where $H[\rho] = E[\rho] - \mu N[\rho]$ and μ is an undetermined-Lagrange-multiplier. If one defines[51]

$$\bar{\rho}(\underline{r}) = \rho(\underline{r}) + \sigma\eta(\underline{r}) \qquad \text{(III-41)}$$

one can express the stationary condition in the form

$$\left.\frac{dH[\bar{\rho}]}{d\sigma}\right|_{\sigma=0} = \int \left.\frac{dh[\bar{\rho}]}{d\sigma}\right|_{\sigma=0} d\tau = 0 \qquad \text{(III-42)}$$

where

$$h(\bar{\rho},\bar{\rho}_x,\bar{\rho}_y,\bar{\rho}_z) = \xi(\bar{\rho},\bar{\rho}_x,\bar{\rho}_y,\bar{\rho}_z) - \mu\bar{\rho} \qquad \text{(III-43)}$$

In this equation we have indicated, in view of Eq.(III-38) the explicit dependence of h and ξ on the derivatives of $\bar{\rho}$ with respect to x, y and z, namely, $\bar{\rho}_x$, $\bar{\rho}_y$ and $\bar{\rho}_z$.

Thus, using Eqs.(III-43) and (III-42) one obtains that

$$\int \left.\left\{\left[\frac{\partial\xi}{\partial\bar{\rho}} - \nabla\cdot\nabla_{\bar{\rho}}\xi - \mu\right]\eta + \nabla\cdot\eta\nabla_{\bar{\rho}}\xi\right\}\right|_{\sigma=0} d\tau = 0 \qquad \text{(III-44)}$$

where

$$\nabla_{\bar{\rho}} = \hat{\imath}\,\frac{\partial}{\partial\bar{\rho}_x} + \hat{\jmath}\,\frac{\partial}{\partial\bar{\rho}_y} + \hat{k}\,\frac{\partial}{\partial\bar{\rho}_z} \qquad \text{(III-45)}$$

and using Green's theorem, Eq.(III-44) may be expressed as

$$\int \left.\left\{\left[\frac{\partial\xi}{\partial\bar{\rho}} - \nabla\cdot\nabla_{\bar{\rho}}\xi - \mu\right]\eta\right\}\right|_{\sigma=0} d\tau + \int_s \eta\nabla_{\bar{\rho}}\xi\cdot d\underline{s} = 0 \qquad \text{(III-46)}$$

The second integral vanishes over the closed surface located at infinity for a well-behaved function, which is the present case, and the condition for the first integral to be zero for an arbitrary value of $\eta(\underline{r})$ is that

$$\left\{\frac{\partial\xi}{\partial\bar{\rho}} - \nabla\cdot\nabla_{\bar{\rho}}\xi - \mu\right\}_{\bar{\rho}=\rho} = 0 \qquad \text{(III-47)}$$

which is precisely the Euler equation.

Thus, substituting Eq.(III-38) into Eq.(III-47) one obtains the Euler equation for such functional,

$$\frac{1}{8}\frac{|\nabla\rho(1)|^2}{\rho^2(1)} - \frac{1}{4}\frac{\nabla^2\rho(1)}{\rho(1)} + \frac{1}{2}(3\pi^2)^{2/3}\,\gamma(N,Z)\rho^{2/3}$$

$$-\frac{Z}{r} + \left(1-\frac{1}{N}\right)\int\frac{\rho(2)}{r_{12}}\,d\tau_2 - \left(\frac{3}{\pi}\right)^{1/3}\delta(N,Z)\rho^{1/3} - \mu = 0 \qquad \text{(III-48)}$$

which may be written in a simpler form by defining

$$u(\underline{r}) = \rho^{1/2}(\underline{r}) \qquad \text{(III-49)}$$

because then, the Euler equation is

$$\Big\{-\frac{1}{2}\nabla^2 + \frac{1}{2}(3\pi^2)^{2/3}\,\gamma(N,Z)u^{4/3}(\underline{r}) - \frac{Z}{r}$$

$$+\left(1-\frac{1}{N}\right)\int\frac{u^2(2)}{r_{12}}\,d\tau_2 - \left(\frac{3}{\pi}\right)^{1/3}\delta(N,Z)u^{2/3}(\underline{r})\Big\}\,u(\underline{r}) = \mu\,u(\underline{r})$$

(III-50)

We can see from this equation that the presence of the Weizsacker term leads to the correct nuclear cusp and long-range behavior of the electronic density.

Now, for $N=2$, $\gamma(2,Z) = 0$ and if one sets $\delta(2,Z) = 0$, Eq.(III-50) becomes identical to the HF equation for two-electron atoms. To test the accuracy of the charge density obtained through this approach for any other atom, we have done a calculation on neon using the Numerov method to integrate Eq.(III-50). The charge density and the radial distribution function are plotted in Fig.3 and compared with the one obtained by the HF and TF methods. One can immediately notice that there is no shell structure as in TF, however, the description specially close to and far away from the nucleus, is greatly improved because of the presence of the Weizsacker term. In addition, the total energy lies closer to the HF value.

The absence of shell structure may be traced back in part, to the term $u^{4/3}(\underline{r}) \sim \rho^{2/3}(\underline{r})$ that represents, in view of Eq.(III-50), for a closed shell atom, an average of $\sum_i \ell i(\ell i + 1)/r^2$. This indicates that although the presence of the corrected TF term in the functional ensures the correct behavior for $Z = N \to \infty$, it may not be adequate for systems far away from this limit, and that one should search for a more realistic $f_{\uparrow\uparrow}(1,2)$ for few-electron atoms if one wishes to obtain more precise charge densities.

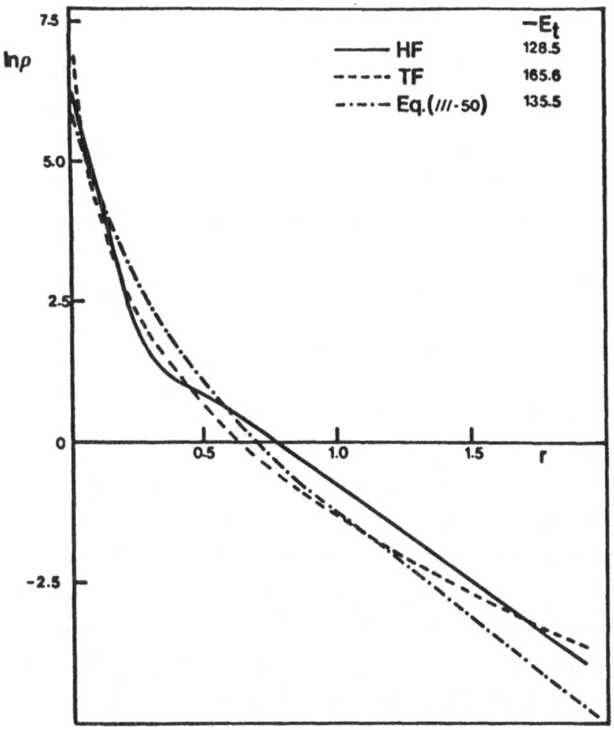

Fig. 3a. Comparison of the charge density obtained by different methods for the Ne atom.

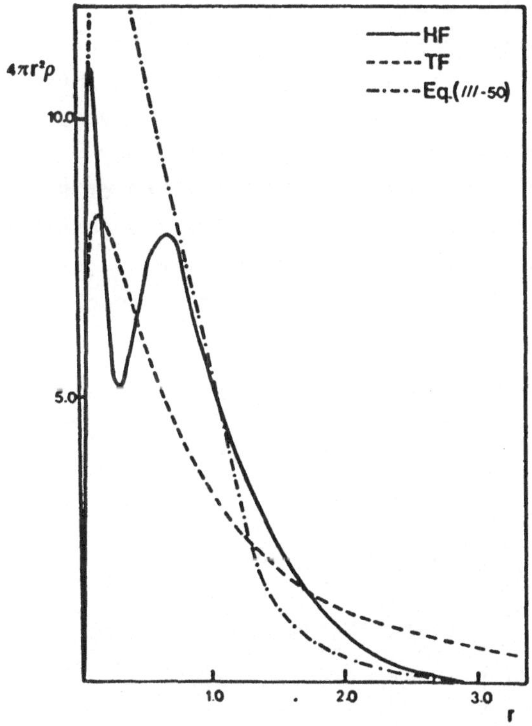

Fig. 3b. Comparison of the radial distribution function obtained by different methods for the Ne atom.

IV. SUMMARY.

In this paper, we have reviewed some aspects of density functional theory in connection with the electron gas model. We have seen that:

1) The Weizsacker term appears as a natural component of the kinetic energy functional when one considers the local variation of the electronic density. Its presence will ensure an adequate Euler equation.
2) The correction to the Weizsacker term is closely related with the exchange energy, since both arise because of the antisymmetry requirement of the wave function.
3) The TF term has been incorporated as a correction, and its presence is important because it has been shown that for $Z = N \to \infty$ is the exact kinetic energy functional.
4) The N-dependence of the kinetic and exchange energies within the electron gas approach, has been established through the corrections to the momentum at the Fermi level, and although in its derivation there were involved several approximations, they were in agreement with the N-dependence obtained through other approaches. Thus it seems that it could be very useful to derive the N-dependence of other properties described through an electron gas model.
5) The total energy can be calculated accurately through the energy formula here derived, when a HF density is used. But, it produces a poor charge density by itself. This situation could be improved by imposing additional constraints in the variation, however, this would lead to a reduction in the generality of the density functional approach. From the arguments used at the end of Sec. III it seems that it would be interesting to analyze the possible presence of terms of the form r^{-2} in the kinetic energy functional[52,53], probably derivable from a non-electron-gas approach.

ACKNOWLEDGEMENTS.

I wish to thank Prof. J. Keller for many useful discussions and E. Ortiz and J. Robles who made major contributions to this work.

References.

1. L.H. Thomas, Proc. Cambridge Philos. Soc. 23, 542 (1927).
2. E. Fermi, Z. Phys. 48, 73 (1928).
3. For a detailed discussion on the Thomas-Fermi model and its various extensions, see: a) P. Gombas, Die Statistische Theorie des Atoms und ihre Anwendung (Springer, Vienna, 1949); b) P. Gombas, Statistische Behandlung des Atoms, in Handbuch der Physik, Bd. 36 (Springer, Vienna, 1956).

4. N.H. March, Adv. Phys. 6, 1 (1957).

5. E.H. Lieb and B. Simon, Phys. Rev. Lett. 31, 681 (1973).

6. E. Teller, Rev. Mod. Phys. 34, 627 (1962).

7. N.L. Baläzs, Phys. Rev. 156, 42 (1967).

8. P.A.M. Dirac, Proc. Cambridge Philos Soc. 26, 376 (1930).

9. C.F. v. Weizsacker, Z. Phys. 96, 431 (1935).

10. D.A. Kirzhnits, Zh. Eksp. Teor. Fiz. 32, 117 (1957); [Sov. Phys. - JETP 5, 64 (1957)]

11. P. Hohenberg and W. Kohn, Phys. Rev. B136, 864 (1964). M. Levy, Proc. Natl. Acad. Sci. USA 76, 6062 (1979).

12. Y.S. Kim and R.G. Gordon, J. Chem. Phys. 60, 1842 (1974).

13. C.C. Shih, Phys. Rev. A14, 919 (1976).

14. W.P. Wang, R.G. Parr, D.R. Murphy and G.A. Henderson, Chem. Phys. Lett. 43, 409 (1976).

15. D.R. Murphy and W.P. Wang, J. Chem. Phys. 72, 429 (1980).

16. C.H. Hodges, Can. J. Phys. 51, 1428 (1973).

17. Y. Tal and R.F.W. Bader, Int. J. Quantum Chem. S12, 153 (1978).

18. D.R. Murphy, Doctoral Dissertation, Dept. of Chemistry, University of North Carolina (1979).

19. F. Herman, J.P. Van Dyke, and I.B. Ortenburger, Phys. Rev. Lett. 22, 807 (1969); F. Herman, I.B. Ortenburger and J.P. Van Dyke, Int. J. Quantum Chem. S3, 827 (1970).

20. J.L. Gázquez, E. Ortiz, and J. Keller, Int. J. Quantum Chem. S13, 377 (1979).

21. C.C. Shih, D.R. Murphy and W.P. Wang, J. Chem. Phys. 73, 1340 (1980).

22. W.P. Wang and R.G. Parr, Phys. Rev. A16, 891 (1977).

23. E. Hernández and J.L. Gázquez, Phys. Rev. A25, 107 (1982).

24. P.K. Acharya, L.J. Bartolotti, S.B. Sears and R.G. Parr, Proc. Natl. Acad. Sci USA 77, 6978 (1980).

25. S.B. Sears, R.G. Parr, and U. Dinur, Isr. J. Chem. 19, 165 (1980).

26. J.L. Gázquez and E.V. Ludeña, Chem. Phys. Lett. 83, 145 (1981).

27. J.L. Gázquez and J. Robles, J. Chem. Phys. 76, 1467 (1982).

28. H. Nakatsuji, Phys. Rev. A14, 41 (1976).

29. L. Cohen and C. Frishberg, J. Chem. Phys. 65, 4234 (1976); L. Cohen and C. Frishberg, Phys. Rev. A13, 927 (1976); C. Frishberg, L. Cohen and P. Blumenau, Int. J. Quantum Chem. S14, 161 (1980).

30. P.W. Payne, Proc. Natl. Acad. Sci. USA 77, 6293 (1980).

31. R. McWeeny, Rev. Mod. Phys. 32, 335 (1960).

32. J.C. Slater, Quantum Theory of Molecules and Solids, (McGraw-Hill, New York), Vol. 4 (1974).

33. J.P. Perdew and A. Zunger, Phys. Rev. B23, 5048 (1981).

34. E. Fermi and E. Amaldi, Mem. Acad. Ital. 6, 117 (1934).

35. R.F.W. Bader and H.J.T. Preston, Int. J. Quantum Chem. 3, 327 (1969).

36. E.V. Ludeña, "On the Nature of the Correction to the Weizsacker Term", submitted for publication in J. Chem. Phys.

37. B.Y. Tong, Phys. Rev. A4, 1375 (1971).

38. J.A. Alonso and L.A. Girifalco, Phys. Rev. B17, 3735 (1978). These authors used a similar approach to the one presented here, but did not consider finiteness corrections.

39. O. Gunnarson, M. Jonson, and B. Lundqvîst, Phys. Lett. A59, 177 (1976).

40. J. Schwinger, Phys. Rev. A22, 1827 (1980).

41. E. Clementi and C. Roetti, At. Data Nucl. Data Tables 14, 177 (1974).

42. I. Lindgren and A. Rosén, Case Studies in Atomic Physics 4, 93 (1974).

43. J.L. Gázquez and E. Ortiz, Chem. Phys. Lett. 77, 186 (1981).

44. S.R. Gadre, L.J. Bartolotti, and N.C. Handy, J. Chem. Phys. 72, 1034 (1980).

45. K. Schwarz, Phys. Rev. B5, 2466 (1972); M. Berrondo and O. Goscinski, Phys. Rev. 184, 10 (1969).

46. M.S. Gopinathan, M.A. Whitehead and R. Bogdanović, Phys. Rev. A14, 1 (1976).

47. J.L. Gázquez and J. Keller, Phys. Rev. A16, 1358 (1977).

48. W. Kohn and L.J. Sham, Phys. Rev. 140, A1133 (1965).

49. J.L. Gázquez, E. Ortiz and J. Robles, to be published.

50. S. Golden, J. Phys. Chem. 83, 1388 (1979); S. Golden, Int. J. Quantum Chem. S14, 135 (1980).

51. I.M. Gelfand and S.V. Fomin, Calculus of Variations (Prentice-Hall, Englewoods Cliffs, New Jersey) 1963.

52. J. Keller, C. Keller and C. Amador, Lecture Notes in Physics (Springer-Verlag) 142, 364 (1981).

53. M. Berrondo and A. Flores-Riveros, J. Chem. Phys. 72, 6299 (1980).

ELECTRON STRUCTURE CALCULATIONS FOR HEAVY ATOMS: A LOCAL DENSITY APPROACH

M.P. Das
International Centre for Theoretical Physics
Trieste, Italy

and

Department of Physics, Sambalpur University
Jyoti Vihar, Sambalpur, 768 017, India
(permanent address)

CONTENTS:

Abstract

We present a brief report outlining a relativistic local density approach to study the electron structure of heavy atoms.

Miramare - Trieste, May 1981

I. MANY BODY IN ONE BODY

An exact treatment of a many particle system has remained to be an unsolved problem in the theoretical physics. Towards a better approximation in the single particle picture Hohenberg and Kohn[1] (HK) have proved two remarkable theorems, which later are generalized to several different situations. These theorems have provided the foundation of a formalism known as "density functional approach." The principal outcome of these theorems is an effective density-dependent one body potential for an interacting many particle system, particularly when the system is in its ground state. This one body potential takes into account the presence of many particles in a statistical manner and is undoubtedly a better representation than that of the mean field Hartree description.

In this brief report we present our attempts to study the electron structure of heavy atoms with the help of the density functional formalism. Because of the presence of a large number of electrons in the atom the electronic density in the core region is very high, such that the electron momentum is comparable with the momentum of a photon. Hence a relativistic treatment of the problem is necessary. Besides vacuum polarization and self-energy corrections of the quantum-electrodynamical in origin would also be desirable. It will not be out of place to mention that the test of the formalism for heavy atoms is a ground work for its application to the complicated systems like molecules and condensed matters involving heavy atoms.

Until now detailed studies of the electronic structure of heavy atoms based on the relativistic theory have been carried out by Dirac-Fock method[2] which are found to be in reasonably good agreement with experimental findings. Unlike the non-relativistic theory parity and total angular momentum (J) are constants of motion in the relativistic framework. So one has to consider several JJ configurations for a system where LS coupling scheme is appropriate which is even true at the Hartree-Fock level. This aspect has given rise to a multi-configurational Dirac-Fock (MCDF) scheme that rests on a very involved analysis. The use of this approach for a complex and large system like a molecule or a solid is a formidable task. Therefore one feels the need for a local approximation-based theory like Slater's X-α method[3] or more correctly a relativistic density functional theory.[4,5] In the recent past several local theory-based calculations[6] have been performed using ad hoc non-relativistic potential in the Dirac equation. But our present scheme differs from these authors in one sense that we are consistent in the relativistic consideration. In Fig. 1 we demonstrate how a relativistic exchange-correlation potential will be different from that of a non-relativistic potential for a mercury atom. In the deep interior of the atom where the electron density is very high, the relativistic contribution makes the potential positive contrary to the non relativistic behavior.[7]

II. LOCAL DENSITY APPROACH

HK variational principle gives a set of self-consistent one particle equations (Dirac equation, being in the present case)

$$H\psi_i \equiv (c\alpha\cdot p+\beta c^2 + V_{eff}[n(\bar{r}),\bar{r}])\psi_i = \varepsilon_i\psi_i \quad (1)$$

$$V_{eff}[n(\bar{r}),\bar{r}] = -\frac{Z}{r} + \int \frac{n(\bar{r}')}{|\bar{r}-\bar{r}'|}\, d\bar{r} + \frac{\delta E_{xc}[n(\bar{r})]}{\delta n(\bar{r})} \quad (2)$$

and

$$n(\bar{r}) = \sum_i^{occ} |\psi_i(\bar{r})|^2 \quad (3)$$

Quantities in (1)-(3) have their usual meanings (Ref. 5). Sum over the occupied states in (3) means positive energy states only. $E_{xc}[n]$ is the exchange correlation energy for which we use local density approximation

$$E_{xc}[n(\bar{r})] = \int d\bar{r}\; \varepsilon_{xc}[n(\bar{r})]n(\bar{r}) \quad (4)$$

$\varepsilon_{xc}[n]$ is the exchange-correlation energy per particle of a homogeneous interacting electron gas which also includes transverse photon-electron interactions. We separate

$$\varepsilon_{xc} = \varepsilon_x + \varepsilon_c \quad . \tag{5}$$

ε_x is the exchange energy corresponding to the second order in perturbation theory arising out of the interaction of a pair of electrons. This contains a contribution from the transverse photon-electron interaction (TPE) known as generalized Breit term (see for example, Brown and Ravenhall[8]). In a approximation where the exchanged photon energy is vanishingly small, TPE term reduces to the Breit result. The well-known result[9] for ε_x is given by

$$\varepsilon_x[n(\bar{r})] = \varepsilon_x^{NR} \left[1 - \frac{3}{2} \left(\frac{\beta(\beta^2+1)^{1/2} - \sin h^{-1}\beta}{\beta^2}\right)^2\right] \tag{6}$$

where ε_x^{NR} is the non-relativistic exchange energy and β being the ratio of the momenta of an electron and a photon in the present context.

It has been found [10,11] that the Breit and the transverse interaction energies for atoms are not too different when these are calculated with the Dirac-Fock orbitals. But we find that transverse interaction energy in LDA for atoms ia bout one and a half times larger than the Breit energy[12] in conformity with the results of MacDonald and Vosko.[4] This disagreement may be a weakness asscociated with LDA. Our Breit exchange potential in LDA is deeper than the transverse-exchange potential (Fig. 2).

The quantity ε_c in Eq. (5) contains the rest of interactions beyond the second order exchange in ε_x. In non-relativistic high density theory one obtains the correlation energy[13] by summing a set of ring diagrams as done by Gell-Mann and Brueckner. But ring sum is not obvious dominant contribution in relativistic theory. Therefore one has to calculate longitudinal and transverse polarization propagators, Q_L and Q_T respectively for the relativistic interacting electron gas in the presence of the transverse electromagnetic field.[9,14] Then ε_c is obtained from

$$\varepsilon_c = \frac{1}{2} \int \frac{d\bar{q}}{(2\pi)^3} \int_{-\infty}^{+\infty} \frac{d\omega}{(2\pi)} [\log(1 + \nu(\bar{q}) Q_L(\bar{q}), \omega) + 2 \log(1 - \frac{8\pi e^2}{q^2+\omega^2} Q_T(\bar{q},\omega))] \tag{7}$$

where $\nu(q)$ is the Fourier transform of the coulomb potential. Our estimates show that for an atom like Hg $\varepsilon_c \cong 3.5$ a.u. which is less than 2 in 10^4 contribution to the total energy.

III. SUMMARY

In this brief report we presented our scheme for the electronic structure calculations through the relativistic local density approach. As said before, this approach is suitable for more complicated systems involving heavy atoms but is not an alternative in the same spirit as multi-configurational Dirac-Fock-Breit theory. One observes an important difference between these two theories that the exchange-correlation (interaction) contributions can be taken into account in a self-consistent manner where MCDF-Breit method is a perturbative approach. There are many interesting properties believed to be associated with relativity.[2] We wish to examine some of them with our present scheme.

ACKNOWLEDGEMENTS

I would like to thank Professor Abdus Salam, the International Atomic Energy Agency and UNESCO for hospitality at the International Centre for Theoretical Physics, Trieste. Thanks are also due to Professor W. Johnson for some useful discussions.

REFERENCES

1. P. Hohenberg and W. Kohn, Phys. Rev. 136B, 864 (1964).

2. J. P. Desclaux, Physica Scripta 21, 436 (1980) and references therein.

3. J. W. D. Connolly in Semiempirical Methods of Electronic Structure Calculations, Part A, ed. by G. A. Segal, Plenum Press, p. 105 (1977).

4. A. H. MacDonald and S. H. Vosko, J. Phys. (C) 12, 2977 (1979), see also A. K. Rajagopal, J. Phys. (C) 11, L943 (1978).

5. M. P. Das, M. V. Ramana and A. K. Rajagopal, Phys. Rev. A22, 9 (1980).

6. See for example, B. Fricke and G. Soff. Atomic Data and Nucl. Data Tables 19, 83 (1977) for atomic calculations. A. Rosen and D. E. Ellis, J. Chem. Phys. 62, 3039 (1975) for molecular calculations and D. Koelling, J. Phys. (Paris) C4-40, 117 (1979) for band calculations.

7. M. P. Das, Phys. Rev. A23, 391 (1981).

8. G. E. Brown and D. G. Ravenhall, Proc. Roy. Soc. A208, 552 (1951).

9. B. Jancovici, Nuovo Cimento 25, 429 (1962).
S. A. Chin, Ann. Physics (N.Y.) 108, 301 (1977).

10. J. B. Mann and W. Johnson, Phys. Rev. A4, 41 (1971).

11. I. P. Grant and B. J. McKenzle, J. Phys. (B) 13, 2671 (1980).

12. M. P. Das, Inter. J. Quantum Chem. 14S, 66 (1980).

13. See, for example, S. H. Vosko, L. Wilk and M. Nusair, Canad. J. Phys. 58, 1280 (1980) and references therein.

14. B. Bezzerides and D. F. Dubois, Ann. Phys. 70, 10 (1972).

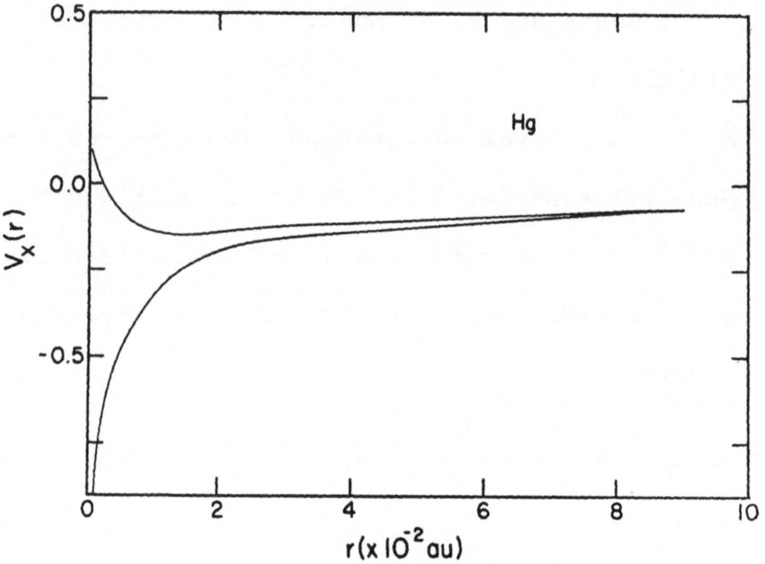

Fig. 1 The self-consistent exchange potential for mercury. The lower curve is for the nonrelativistic X-α potential ($\alpha = 2/3$) while the upper is for the relativistic exchange potential.

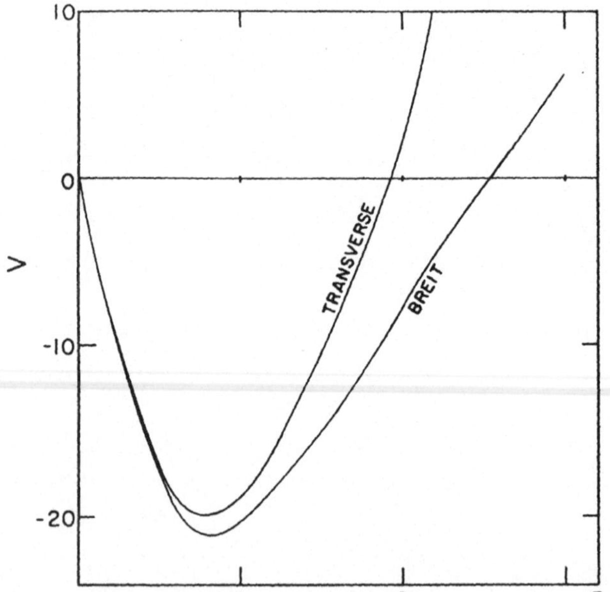

Fig. 2 Exchange potentials including TPE and Breit corrections are shown as function of β.

DENSITY FUNCTIONALS OBTAINED FROM MODELS OF THE ELECTRON FIRST AND SECOND ORDER DENSITY MATRICES

Jaime Keller and Carlos Amador
D.E.Pg., Facultad de Quimica &
Facultad de Estudios Superiores, Cuautitlan
Universidad Nacional Autonoma de Mexico
04510, Mexico, D.F.

CONTENTS:

INTRODUCTION

We review the current use of the free electron gas density matrices model to obtain local electronic density functionals and make it explicit that the only parameter in the theory is the free electron gas parameter $\rho^{1/3}$. For spherically symmetric charge densities, additional terms including $1/r$, $1/r^2$, $\partial/\partial r$ and $\partial^2/\partial r^2$ should be studied to obtain a density functional including symmetry and statistics from the beginning. The results that could be expected are analyzed numerically.

THE DENSITY FUNCTIONAL.

The main objective of the density functional method is to obtain the ground state energy of a many electron system from a knowledge of the electronic density and to understand this relation.

The final goal would be to obtain both the ground state of the many particle system and the density of particles selfconsistently. The basic formalism and foundations of the method are thoroughly discussed in other papers of this volume.

In particular, we want to obtain a unique and continuous functional for the total energy, customarily analyzed in the following form:

a) Kinetic energy

$$E_{kin} = \int d^3r \int d^3r' \delta(r - r') \left[\frac{-\nabla_r^2}{2m} \rho(r,r') \right] = \int d^3r t(r) \tag{1}$$

b) Electron nuclear energy

$$E_{en} = \int d^3r \rho(\underline{r}) V_n(\underline{r}) \tag{2}$$

c) Coulomb part of the electron-electron energy

$$E_{ee} = \frac{e^2}{2} \int d^3r \int d^3r' \frac{\rho(\underline{r})\rho(\underline{r}')}{|\underline{r} - \underline{r}'|} \tag{3}$$

d) Exchange part of the electron-electron energy

$$E_{ex} = - \frac{e^2}{2} \int d^3r \int d^3r' \frac{\rho(r,r')\rho(r',r)}{|r - r'|} = \int d^3r \rho_x(r) \tag{4}$$

e) Correlation part of the electron-electron energy

$$E_{corr} = \frac{e^2}{2} \int d^3r \int d^3r' \frac{\rho^{(2)}(\underline{r}\underline{r}',\underline{r}\underline{r}')}{|\underline{r}-\underline{r}'|} - E_{ee} - E_{ex} \tag{5}$$

Defined by the relations given above, where the constants have the usual meaning, the formulae are also used to define the energy densities, and where $\rho(\underline{r})$ and $\rho(r,r')$ are the one and two particle densities.

A most useful starting point has been in the past, the definition of operators, their eigenvalues and spectral decomposition as follows. A complete set of eigenstates of the energy is defined by

$$(\hat{t} + \hat{V}_{eff})\,|\alpha\rangle = \varepsilon_\alpha|\alpha\rangle \tag{6}$$

and is used to expand the two-particle densities, knowing the occupation numbers n_α as

$$\langle\underline{r}|\hat{\rho}|\underline{r}'\rangle = \rho(\underline{r},\underline{r}') = \sum n_\alpha \langle\underline{r}|\alpha\rangle\langle\alpha|\underline{r}'\rangle \tag{7}$$

The occupation numbers, in an operator representation, are

$$n_\alpha \rightarrow \theta(\varepsilon_F - \varepsilon_\alpha) \rightarrow \theta(\varepsilon_F - \hat{t} - \hat{V}_{eff}) \tag{8}$$

to obtain, finally

$$\rho(\underline{r},\underline{r}') = \langle\underline{r}|\theta(\varepsilon_f - \hat{t} - \hat{V}_{eff})|\underline{r}'\rangle \tag{9}$$

Historically, and because the momentum spectral analysis of the expectation values of operators is so convenient, the starting point has been the energy degenerated eigenstates of the free electron gas. This allows a convenient representation of the density operator as follows. First a Fermi kinetic energy operator is defined

$$\hat{E}_F = \varepsilon_F - \hat{V}_{eff} \tag{10}$$

and replaced in (9) to obtain (θ being the Heaviside operator)

$$\hat{\rho} = \theta(\hat{E}_F - \hat{t}) \tag{11}$$

From it, the expectation values of the one-electron operators are easily evaluated with the spectral decomposition of the two-particle density

$$\rho(\underline{r},\underline{r}') = 2\int d^3k \langle\underline{r}|\theta(\hat{E}_F - t)|\underline{k}\rangle\langle\underline{k}|\underline{r}'\rangle \qquad \{|\underline{k}\rangle\} \text{ complete} \tag{12}$$

The set $|\underline{k}|$ being complete. This intermediate set of eigenfunctions has the great advantage of being simultaneous eigenfuctions of the energy

$$\hat{t}|\underline{k}\rangle = \frac{k^2}{2m}|\underline{k}\rangle \tag{13}$$

with

$$\rho \approx \rho_o(r,r') = \frac{2}{(2\pi)^3}\int d^3k\,\theta\left(\hat{E}_F(r) - \frac{k^2}{2m}\right) e^{i\underline{k}\cdot(\underline{r}-\underline{r}')} \tag{14}$$

A local Fermi momentum is defined

$$K_F(\underline{r}) = \sqrt{2m\,E_F(r)} \tag{15}$$

such that (the $j_\ell(x)$ is the spherical Bessel function)

$$\rho_o(\underline{r},\underline{r}') = \frac{K_F^2}{\pi^2} j_1(K_F y)\frac{1}{y} \qquad \underline{y} = \underline{r} - \underline{r}' \tag{16}$$

is the most widely employed approximation to the two-particle density with the limit $y \to 0$ for the relationship between the local Fermi momentum and density

$$\rho_o(r) = \frac{1}{3\pi^2} K_F^3(r) \tag{17}$$

Higher order approximations have been devised to correct for the fact that the local density changes from place to place in actual systems. Here we present a quick review of the Kirzhnits method [1], using the Laplace transforms of a function of the sum of two non-commuting operators $\hat{a}$ and $\hat{b}$

$$f(\hat{a}+\hat{b})|a\rangle = \int d\lambda\; e(\lambda)e^{\lambda(\hat{a}+\hat{b})}|a\rangle \tag{18}$$

where the exponential is assumed to be expanded as

$$\exp\left[\lambda(\hat{a}+\hat{b})\right] = \exp\left[\lambda b\right] K(\lambda)\exp\left[\lambda\hat{a}\right] \tag{19}$$

where the operator $K(\lambda)$ should obey the following formula and boundary condition

$$\frac{d\hat{K}}{d\lambda} = \left[\hat{a},\hat{K}\right] + \exp(-\lambda b)\left[\hat{a}, \exp(\lambda b)\right]\hat{K}(\lambda) \qquad \hat{K}(0) = 1 \tag{20}$$

Usually $K(\lambda)$ is assumed to be obtained from a series

$$\hat{K}(\lambda) = \sum_{n=o}^{\infty} \lambda^n \hat{O}_n \tag{21}$$

with the following definitions and boundary condition for the operators $\hat{O}$

$$\hat{O}_n = \frac{1}{n}\left(\left[\hat{a}\,,\hat{O}_{n-1}\right] + \sum_{i=1}^{n} \hat{C}_i \hat{O}_{n-1-i}\right)$$

$$\hat{C}_i = \frac{(-1)^i}{i!}\left[\hat{b},\left[\hat{b},\left[\ \ldots\ \left[b,a\right]\right]\right]\right]$$

$$\hat{O}_o = 1 \qquad \hat{O}_1 = 0 \tag{22}$$

If we know the eigenstates of operator a, we can write for (19)

$$\begin{aligned}\exp\left[\lambda(\hat{a}+\hat{b})\right]|a\rangle &= \exp(\lambda\hat{\underline{b}})\,\hat{K}(\lambda)\exp(\lambda\hat{a})\,|a\rangle \\ &= \exp\left[\lambda(a+\hat{b}\right]\hat{\hat{K}}(\lambda)\,|a\rangle\end{aligned} \tag{23}$$

and introducing the formal derivative of f(Q)

$$f^{(n)}(\hat{Q}) = \frac{d^n f(Q)}{d\hat{Q}^n} = \int d\lambda\, C(\lambda)\,\lambda^n \exp\left[\lambda\hat{Q}\right] \tag{24}$$

obtain the series expansion

$$f(\hat{a}+\hat{b})\,|a\rangle = \sum_{n=o}^{\infty} f^{(n)}(a+\hat{b})\hat{O}_n\,|a\rangle \tag{25}$$

All this will be used in connection with (11) and (14) with the following substitutions

$$f \to \theta \qquad \hat{a} \to -\hat{t} \qquad \hat{b} \to \hat{E}_F$$

and a local representation of E_F. Here the commutators of $\hat{E}_F$ with the free electron Hamiltonian $(-\hat{a})$ are the successive derivatives of the Fermi energy. This is the basis of the well known expansions

$$\rho(\underline{r}) = \frac{K_F^3(\underline{r})}{3\pi^2} + \frac{1}{24\pi^2}\frac{\nabla^2 K_F^2}{K_F} - \frac{1}{96\pi^2}\frac{(\nabla K_F^2)^2}{K_F^3} + \ldots$$

$$\tau(\underline{r}) = \frac{K_F^5(\underline{r})}{10m\pi^2} - \frac{1}{48m\pi^2}K_F\nabla^2 K_F^2 - \frac{1}{64m\pi^2}\frac{(\nabla K_F^2)^2}{K_F} + \ldots$$

$$\rho_x(\underline{r}) = -\frac{e^2}{4\pi^3}K_F^4(\underline{r}) - \frac{e^2}{576\pi^3}\frac{(\nabla K_F^2)^2}{K_F^2} + \ldots \tag{26}$$

where, usually, only the first order term in the density operator is kept

$$\rho_o(r) = \frac{1}{3\pi^2}K_F^3(r) + \ldots \tag{27}$$

to obtain the working formulae

$$\tau(\underline{r}) = c_1 = \frac{3}{10m}(3\pi^2)^{2/3} + C_2 = \frac{1}{72m}$$

$$\rho_x(\underline{r}) = -C_3 = e^2\,\frac{3}{4}\left(\frac{3}{\pi}\right)^{1/3} - C_4 = \frac{7e^2}{432\pi(3\pi^2)^{1/3}} \tag{28}$$

The same procedure has allowed the treatment of the relativistic problem, the work of Dreizler and Gross [2] starts from a relativistic ground state total energy, within the single particle approximation, as

$$E_o^{(rel)} = \mathrm{tr}\left[\int d^3r\int d^3r'\ \delta(\underline{r}-r')(-i\underline{\alpha}\cdot\underline{\nabla}_r+\beta m)\rho(\underline{r},\underline{r}') + \int d^3r\rho(\underline{r},\underline{r})\ v_n(\underline{r}) + \frac{1}{2}\int d^3r\int d^3r'\rho(\underline{r},\underline{r})V(\underline{r},\underline{r}')\rho(\underline{r}',\underline{r}') - \frac{1}{2}\int d^3r\int d^3r'\rho(\underline{r},\underline{r}')V(\underline{r},\underline{r}')\rho(\underline{r}',\underline{r})\right]. \tag{29}$$

Where the particle-particle interaction should be given by the series

$$V(\underline{r},\underline{r}') = \frac{e^2}{|\underline{r}-\underline{r}'|} - \frac{e^2}{2}\left[\frac{\alpha\cdot\alpha'}{|\underline{r}-\underline{r}'|} + \frac{\alpha\cdot(\underline{r}-\underline{r}')\alpha'\cdot(\underline{r}-\underline{r}')}{|\underline{r}-\underline{r}'|^3}\right] + \ldots \tag{30}$$

although they restricted themselves to the first term. At the end of this paper we will comment on the transverse interaction.

Now they define

$$\hat{t} = \underline{\hat{\alpha}}\cdot\underline{\hat{p}} + \hat{\beta}m \tag{31}$$

and

$$\hat{\rho}_{rel} = \theta(\varepsilon_F - \hat{h}_{eff}) - \theta(-m - \hat{h}_{eff}) \tag{32}$$

with

$$\hat{h}_{eff} = \underline{\hat{\alpha}}\cdot\underline{\hat{\alpha}} + \hat{\beta}m + \hat{v}_{eff} \tag{33}$$

to obtain

$$\rho(\underline{r},\underline{r}') = \langle\underline{r}|\hat{\rho}_{rel}|\underline{r}'\rangle = \sum_{sT}\int d^3k\ \langle\underline{r}|\theta(E_F(\underline{r}) - \hat{t})|\underline{k}sT\rangle\langle\underline{k}sT|\underline{r}'\rangle \quad \sum_{sT}\int d^3k\ \langle\underline{r}|\theta(G_F(\underline{r}) - \hat{t})|\underline{k}sT\rangle\langle\underline{k}sT|\underline{r}'\rangle$$

with $\hat{E}_F = \varepsilon_F - \hat{v}_{eff}$

$$G_F = -m - \hat{v}_{eff} \tag{34}$$

where the total density is expressed as the difference between particles and "holes"

$$\rho_o(\underline{r}) = \rho_1(\underline{r}) - \rho_2(\underline{r})$$

with

$$\rho_1(\underline{r}) = \frac{1}{3} K_F^3 \ , \ K_F(v_{eff}) = \left[E_F^2 - m^2\right]^{1/2} = \left[(\varepsilon_F - v_{eff})^2 - m^2\right]^{1/2}$$

and

$$\rho_2(\underline{r}) = \frac{1}{3\pi^2} q_F^3 \ , \ q_F(v_{eff}) = \begin{cases} \left[G_F^2 - m^2\right]^{1/2} = \left[(-m - v_{eff}) - m\right]^{1/2} \\ 0 \qquad \text{if } v_{eff} > -2m \end{cases}$$

(35)

with the final result for the kinetic energy density

$$\tau(\underline{r}) = \tau_o(x) + \frac{1}{72m} B(x) \frac{(\nabla\rho_1)^2}{\rho_1} - \tau_o(y) - \frac{1}{72m} B(y) \frac{(\nabla\rho_2)^2}{\rho_2}$$

with

$$B(Z) = \frac{1}{\sqrt{1+Z^2}} + \frac{2Z}{1+Z^2} \text{Arsh } Z \tag{36}$$

The quantities x and y are given by

$$x := K_F/m = (3\pi^2\rho_1)^{1/3}/m \tag{37}$$

and

$$y := q_F/m = (3\pi^2\rho_2)^{1/3}/m. \tag{38}$$

Dreizler and Gross comment on the exchange energy density which will be corrected by ρ_2, although no working formula has been presented yet.

DENSITY FUNCTIONALS FOR ELECTRONIC EXCHANGE.

The expression for the exchange potential in terms of the electronic density and the pair correlation function can be modeled in several different ways. In this section we present the main forms used until now and propose some new exchange potentials which should have a local behavior in better agreement with Hartree-Fock average exchange potential.

Since the early forms of the Thomas-Fermi approximation and statistical exchange the actual Hartree-Fock, non-local exchange terms have been replaced by an average over electronic states:

$$U_{xHF}(1) = -\sum_{i\uparrow,j\uparrow} n_i n_j \int u_i^*(1) u_j^*(u_j(1) u_i(2) \frac{1}{r_{12}} dv_2 \left(\sum_{k\uparrow} n_k u_k^*(1) u_k(1)\right)^{-1} \tag{39}$$

or for the total exchange energy

$$E_{x\,HF} = E_{x\,HF\uparrow} + E_{x\,HF\downarrow} = \frac{1}{2}\int \left[\rho^{\uparrow}(1)U_{x\,HF^{\uparrow}}(1) + \rho^{\downarrow}(1)U_{x\,HF^{\downarrow}}(1)\right] dv_1 \quad (40)$$

These approximations give different potentials for different spins. It is much easier to analyze the different approaches used until now and to search for new ones if the electron-pair correlation functions $C^{\uparrow}(1,2) = \rho^{\uparrow}(r,r')$ are used instead of the two electrons density matrices $\Gamma^{\uparrow}(1,2)$

$$C^{\uparrow}(1,2) \equiv \frac{\Gamma^{\uparrow}(1,2)}{\rho^{\uparrow}(1)\rho^{\uparrow}(2)} - 1 \quad (41)$$

the exchange energy E_x being then defined as

$$E_{x\,HF} = \frac{1}{2}\int \rho^{\uparrow}(1) \left(\int \frac{1}{r_{12}} \rho^{\uparrow}(2)C^{\uparrow}(1,2)dv_2\right) dv_1$$

$$+ \frac{1}{2}\int \rho^{\downarrow}(1) \left(\int \frac{1}{r_{12}} \rho^{\downarrow}(2)C^{\downarrow}(1,2)dv_2\right) dv_1 \quad (42)$$

or the exchange energy density $U_x(1)$

$$U_x(1) = \int \rho^{\uparrow}(2)C^{\uparrow}(1,2) \frac{1}{r_{12}} dv_2 \quad (43)$$

From these expressions (42, 43) the density functionals for exchange energy and potentials can be obtained. The different possibilities are:

a) $C^{\uparrow}(1,2)$ is modeled and $\rho(2)$ is taken to be the actual density in a self-consistent calculation.

b) $C^{\uparrow}(1,2)$ and $\rho(2)$ are modeled.

c) The result of the integral (43) is modeled directly.

In any one of the three approximations there are some sum rules to be fulfilled which help to find the parameters of the model.

First the total charge $n^{\uparrow}$ should be conserved

$$\int \rho^{\uparrow}(2)\, dr_2 = n^{\uparrow} \quad (44)$$

and, second, the exchange charge should be

$$\int \rho^{\uparrow}(2)C^{\uparrow}(1,2)dr_2 = -1. \quad (45)$$

The most popular models used have been of type b) above.

Type a) The use of the free electron gas pair correlation formula

$$C^{\uparrow}(1,2) \approx C^{\uparrow}_{FE}(1,2,y) = -\frac{9}{2}\left(\frac{\text{sen } y - y\cos y}{y^3}\right) \quad (46)$$

using

$$y = r_{12}\left[3\pi^2\rho^{\uparrow}(1)\right]^{1/2} \quad (47)$$

and the actual charge density by Alonso and Girifalco [3] and Gunnarson, Jonson and Lundqvist [4] or the use of a model pair correlation funtion function by Keller, Keller and Amador [5] ,

$$C^{\uparrow}(1,2) \approx C_M^{\uparrow}(1,2) = -e^{-br_{12}} \tag{48}$$

In all cases the sum rule (45) is used to fix the parameters of $C^{\uparrow}$.

Of type b) above

b-1) The original Dirac-Slater [6] approximation

$$\rho(2) \approx \rho(1) \tag{49}$$

$$C^{\uparrow}(1,2) \approx C_{FE}^{\uparrow}(1,2) \tag{50}$$

b-2) A geometrical approach to the Fermi hole with an improved total number of electrons boundary condition by Gopinathan, Whitehead and Bogdanovic [7]

$$\rho(2) \approx \rho(1) \tag{51}$$

$$C^{\uparrow}(1,2) \approx C_{GWB}^{\uparrow}(1,2) \tag{52}$$

and the

b-3) Model Fermi hole with improved boundary condition for the total number of electrons of Gázquez and Keller (8)

$$C_{GK}^{\uparrow} = C_1 H(1,2) + C_2 ; \quad H(r) = \exp\left(\frac{-ar}{r_f}\right)\left[1 + \frac{br}{r_f} + c\left(\frac{r}{r_f}\right)^2\right] \tag{53}$$

b-4) Both the Slater exchange and the Gázquez and Keller exchange have been improved by including the gradient terms in the approximation for the charge density [9]

$$\rho(2) \approx \rho(1) + \left.\frac{\partial\rho(r)}{\partial r}\right|_1 r_{12} \tag{54}$$

and

$$C^{\uparrow}(1,2) = \begin{cases} C_{FE}^{\uparrow}(1,2) \\ C_{GK}^{\uparrow}(1,2) \end{cases} \tag{55}$$

finally

b-5) An attempt to use a more realistic charge density within the model has been analyzed by Keller, Keller and Amador. For example at each radius r_1 the approximation

$$\rho(2) \approx A_1 e^{-b_1 r_2} \tag{56}$$

has been used, with A_1 and b_1 given by

$$-b_1 = \frac{\partial}{\partial r} \ln \rho(r)\Big|_{r_1} \tag{57}$$

$$\rho(1) = A_1 e^{-b_1 r_1} \tag{58}$$

the use of higher derivatives allows the possibility of including more parameters in the proposed function for $\rho(2)$. With this approach to $\rho(2)$ some different approximations for the pair-correlation function can be used, basically

$$\rho(2) \approx A_1 r^{-b_1 r_2} \tag{59}$$

$$C^{\uparrow}(1,2) \approx C^{\uparrow}_{FEy}(1,2) \tag{60}$$

In relation to models type c) we may quote the work of Berrondo and Goscinski [10] which from considerations of the two-electron density matrices derived the expression

$$U_x^{BG}(1) = \frac{\pi A}{\eta^3 r}\left[1 - (1 + \eta r)e^{-2\eta r}\right] dr \quad \text{with} \quad \rho(r) = A\, e^{-2\eta r} \tag{61}$$

and, as we want to suggest in the present paper, the possibility of using a function $U_x(r)$ with the boundary conditions of being finite at the origin and to behave like $1/r$ for large r

$$U_x(r) = -\, f(r)\, \frac{1}{r} \tag{62}$$

$$\lim_{r\to\infty} f(r) = 1 \tag{63}$$

$$\lim_{r\to 0} f(r) = ar; \quad a = \frac{3}{2}\left(\frac{3}{\pi}\right)^{1/3} \rho(o)^{1/3} \tag{64}$$

a suitable form for $f(r)$ is

$$f(r) = 1 - e^{-ar} \tag{65}$$

In [5] was found that the total exchange energy E_x is well approximated by an r independent constant value

$$f(r) = A - b\, Z \tag{66}$$

Z being the atomic number.

KINETIC ENERGY AND COULOMB ELECTRON-ELECTRON INTERACTION DENSITY FUNCTIONALS FOR SPHERICALLY SYMMETRIC PROBLEMS.

The well known Thomas-Fermi theory with its subsequent development, provides a local density functional in terms of powers of the free electron gas parameter $\rho^{1/3}$. For spherically symmetric charge densities, additional terms including $1/r$, $1/r^2$, $\partial/\partial r$ and $\partial^2/\partial r$ should be studies and we analize the way they arise from the study of charge densities and the results that could be expected with their use.

We have seen that the derivation of electronic density functionals for the total energy is made from both physical considerations and models of the electron charge density. For the free electron gas case there is only one parameter in the theory: the density ρ, then the dimensional quantities that should be formed (x^{-1} and x^{-2}) can be obtained from $\rho^{1/3}$ and $\rho^{2/3}$ for the potential and kinetic energy parts respectively. Corrections to these terms are obtained either from dimensionless functionals of ρ or from an appropriate use of the operator $\partial/\partial x$ or $\partial^2/\partial x^2$. For spherically symmetric charge densities (a sum over angular momenta components) the terms allowed in the theory are $\rho^{1/3}$, $\rho^{2/3}$, $\partial/\partial r$ and $\partial^2/\partial r^2$, besides the obvious 1/4 and $1/r^2$.

First we will show how this can be done with illustrations for (spherically symmetric) atomic charge densities where the possible kinetic energy functional for the central symmetric problem is

$$E = \int \left\{ -\frac{1}{4}\, \partial^2/\partial r^2 - \frac{1}{r}\, \partial/\partial r + A/r^2 + 0.5\, \partial^2 \ln\rho/\partial r^2 \right\} \rho(r) 4\pi r^2 dr \qquad (67)$$

It will be found that A should not be strictly constant but depend on Z. The use of the value A = - .0519 + .0041 Z gives results which agree with the total kinetic energy within 3% for the first 45 atoms of the periodic table. But as the difference with the Hartree-Fock total kinetic energy depends smoothly with Z, an optimized value of A could be used for a given range of atomic numbers to obtain the best results in practice.

In the last part of this section we analyse the Coulomb electron-electron interaction.

The Thomas-Fermi theory performs a free electron gas analysis of the various energy terms. It could therefore be possible to introduce the equivalent information for other systems in the construction of the density functionals. This is made here for spherically symmetric atoms. The procedure can be improved systematically and extended to molecules and condensed matter.

To develop a density functional we shall start from the Schrödinger equation for an atom (in Rydberg units)

$$\sum_i \left\{ -\nabla_i^2 - \frac{2Z}{r_i} + \sum_{j\neq i} \frac{1}{r_{ij}} \right\} \psi(\{\underline{r}_i\}) = E\psi(\{\underline{r}_i\}) \tag{68}$$

and introduce a Hartree electronic wave function for an atom ψ_H

$$\psi_H(\{\underline{r}_i\}) = \prod_i \phi_i(\underline{r}_i) \tag{69}$$

in equation (68) this can be done if an exchange-correlation effective potential is introduced.

The choice of the set of monoelectronic wave functions ϕ_i contains the long known rules to find either the lowest energy state for an atom or a fixed <u>ad hoc</u> electronic configuration.

When electron gas theory is used the total kinetic energy is directly found, but here we will split it into radial kinetic energy and angular kinetic contributions, where the expectation value of the angular part of the kinetic energy

$$K_\Omega = \sum_i \ell_i(\ell_i + 1) < \frac{1}{r_i^2} > \tag{70}$$

The main idea of this section is that if an electronic configuration and a form for the monoelectronic functions ϕ_i are given or, equivalently, of the density matrices, an equation for the total energy can be written explicitly. The resulting functional can then be parametrized <u>a posteriori</u> to optimize it.

Let us assume in the present example that (neglecting orthogonality for different n of a given ℓ) each ϕ_i is of the type

$$\phi_i(\underline{r}_i) = A r^{\ell_i} e^{-ar_i} Y_{\ell_i m_i}(\hat{r}_i) \tag{71}$$

A and a depend on n_i and ℓ_i.

Then the n-electron wave function (69) can be substituted in (68). The angular integrations are performed, the Hartree-Coulomb potential is introduced together with the exchange-correlation energy density $\varepsilon_{xc}(r)$, and the result is multiplied on the left by the wave function. The result of these manipulations is

$$\sum_i \left\{ -\frac{1}{4}\left\{\frac{\partial^2}{\partial r^2} + \frac{4}{r}\frac{\partial}{\partial r}\right\} + \frac{\ell_i}{2r^2} + \frac{\ell_i(\ell_i+1)}{r^2} - \frac{2Z}{r} + \int_0^r \frac{\rho(r')}{r} 4\pi r'^2 dr' + \right.$$
$$\left. + \int_r^\infty \rho(r') 4\pi r' dr' + \varepsilon_{xc}(r) \right\} |\psi|^2 = E|\psi|^2 \tag{72}$$

Finally, from this expression, a density functional can be obtained averaging over the ℓ_i.

For each summation in (71) the integral over the j≠i has to be made; this is straightforward for the first and the last four terms. If a suitable A could be found, the final functional would be

$$\int \left\{ -\frac{1}{4}\left[\frac{\partial^2}{\partial^2} + \frac{4}{r}\frac{\partial}{\partial r}\right] + C\frac{1}{r^2} - \frac{2Z}{r} + \varepsilon_{coul}(r) + \varepsilon_{xc}(r) \right\} \rho(r)d^3r = E \qquad (73)$$

Use of the one-particle density matrix $\rho(r)$ is allowed by the fact that the local electron-electron potential $\varepsilon_{cou}(r) + \varepsilon_{xc}(r)$ has been introduced.

In equation (73) the (z dependent)

$$C = \overline{\ell(\ell + 3/2)} = \left(\sum_i \langle \ell_i(\ell_i + 3/2) r_i^{-2} \rangle\right) \left(\sum_i \langle r_i^{-2} \rangle\right)^{-1} \qquad (74)$$

stands only for the form (70) of the wave function, a more general (orthogonal) set would not allow this definition. The term $\ell_i/2r^2$ can also be obtained from (70) using $\partial^2 \ln\rho/\partial r^2$ as an extra term in (72).

The functional (72) should be used with a density which avoids the inaccuracies introduced. This can be accomplished if the density is constructed as

$$\rho(r) = \sum_i \phi_i^2(r) \qquad (75)$$

with improved forms for Φ_i (spherical averages performed to avoid cross terms, as usual in practice for density functional theory). The parameters used to construct the final ϕ can, otherwise, be optimized from the functional itself. The free electron gas density can be expressed as a sum over spherical harmonics and used with (72).

The Coulomb, exchange and correlation energy densities may be given in a local density functional approximation also. The model of the exchange and correlation part of the potential can be extended to the Coulombic term

$$V_{coul}(r) = \frac{4}{3} A\rho^{1/3}(r) = \frac{\partial \varepsilon_{coul}(r)}{\partial \rho(r)} \qquad (76)$$

if a form is given for the electron gas pair-correlation function. A suitable first approximation for the second-order density matrix could be

$$\Pi(1,2) = \rho^2(1) \left\{ e^{-r_{12}/r_o} - f_{xc}(1,2) \right\} \quad , r_{12} \leq r_o ; \tag{77}$$

with r_o adjusted in such a way that the total charge n (in units of the electron charge) is

$$\rho(1) \int_0^{r_{max}} e^{-r_{12}/r_o} \, 4\pi r_{12}^2 dr_{12} = n \tag{78}$$

$r_{max} = mr_o$, m is an integer we fix so that the model obeys the virial theorem, and $f_{xc}(1,2)$ is described in references [8,9] as in (53), including the spin polarized case. The energy density $\varepsilon_{cou}(r)$ resulting from the Coulombic potetial of the electron gas is

$$\varepsilon_{cou}(r) = 1.7921 \, z^{2/3} \, \rho^{4/3} = D\rho^{4/3} \tag{79}$$

A similar approach for the Coulomb energy was introduced by Parr, Gadre, Bartolotti and Handy [11] in the Hohenberg-Kohn-Sham formalism [12].

The density (75) can be fixed or given a form with parameters which will be optimized by conventional methods.

In the case of molecular and solid state physics, the functionals here discussed can be used directly in the renormalized atom approach, the one center and some cellular methods, allowing for straightforward evaluation of total energies.

We should mention that there are alternative forms of writing the functional (72). The numerical analysis which follows should be useful in this respect.

Table 1 shows the contributions to the kinetic energy obtained from (near) Hartree-Fock charge densities using (67) and, as a comparison, the Hartree-Fock values (HF) and the Thomas-Fermi (TF) values ($E_k = \frac{3}{5}(3\pi^2)^{2/3} \int \rho^{5/3} d\tau$). The first (gradients) term in (67) is insufficient but good for light atoms. The $\partial^2 \ln\rho/\partial r^2$ (logarithmic) term brings the results closer to HF for medium atoms but overestimates the kinetic energy for the light atoms. The third term contains then possitive as well as negative contributions. Equation (67)gives kinetic energies very close to HF a r.m.s. percentage error of 1.8. This is better than Thomas-Fermi plus the original Weizacker correction for non-homogeneity but not as good as improved Thomas-Fermi functionals [13].

TABLE I. KINETIC ENERGY COMPONENTS FOR HARTREE-FOCK DENSITIES.

Z	(1)	(2)	(3)	% error	HF	TF
2	5.5075	6.2986	5.8169	- .0260	5.9720	5.121
3	15.1450	16.8082	15.6083	.0337	15.1000	13.359
4	29.0181	31.9527	29.8923	.0216	29.2593	26.256
5	47.1706	53.1643	50.2025	.0218	49.1310	43.930
6	69.9268	80.9363	77.1199	.0219	75.4653	67.296
11	257.7361	328.1413	324.6415	.0026	323.8063	297.56
12	311.3269	399.2921	397.6140	- .0041	399.2643	368.00
13	370.4601	478.4927	479.5279	- .0087	483.7402	446.74
15	505.5525	661.1085	670.7906	- .0156	681.4457	631.08
16	581.6776	765.0021	780.8960	- .0178	795.0112	737.22
17	663.6967	877.5827	901.1391	- .0194	918.9805	853.36
20	946.6783	1268.4628	1325.2371	- .0209	1353.593	1260.14
25	1544.3070	2115.9185	2271.3919	- .0126	2300.4271	2143.26
28	1985.1040	2754.1175	3002.3203	- .0042	3014.8838	2816.00
30	2316.4070	3238.7435	3565.9148	.0025	3557.1981	3331.44
33	2875.5720	4059.5875	4535.5042	.0146	4470.0241	4197.58
35	3291.5430	4675.2415	5273.1244	.0210	5164.5627	4840.38

$(1) = -\frac{1}{4}\,\partial^2/\partial r^2 - \frac{1}{r}\,\partial/\partial r$ $\quad (2) = (1) + 0.5\,\partial^2 \ln\rho/\partial r^2$ $\quad (3) = (2) + A\left(\frac{1}{r^2}\right)$

$A = -0.0519 + 0.0041Z$ $\quad$ r.m.s. = 0.0181

Figure 1 shows the above results more clearly and suggests that the scaling of any one of the terms (alone or with a second one) could be useful. Figure 2 shows the value of A in equation (67)that should be used, if the "logarithmic" therm is suppressed, to obtain the HF kinetic energies.

On the other hand the total HF kinetic energies E_k could have been obtained from the parametrized formulae

$$E_k = 0.7722\, Z^{1/5} \left< -\frac{1}{4}\frac{\partial^2}{\partial r^2} - \frac{1}{r}\frac{\partial}{\partial r} \right> \text{ (Ry)} \tag{80}$$

$$E_k = 0.3861\, Z^{1/5} \left< 1/r^2 \right> \text{ (Ry)} \tag{81}$$

(the first coefficient is the double of the second!)

These formulae should be useful to obtain other type of relations, for example from the Parr and Gadre relation [14].

$$E_k = - E_k = - 1.0398\ Z^{2.3947}\ \text{(Ry)} \qquad (82)$$

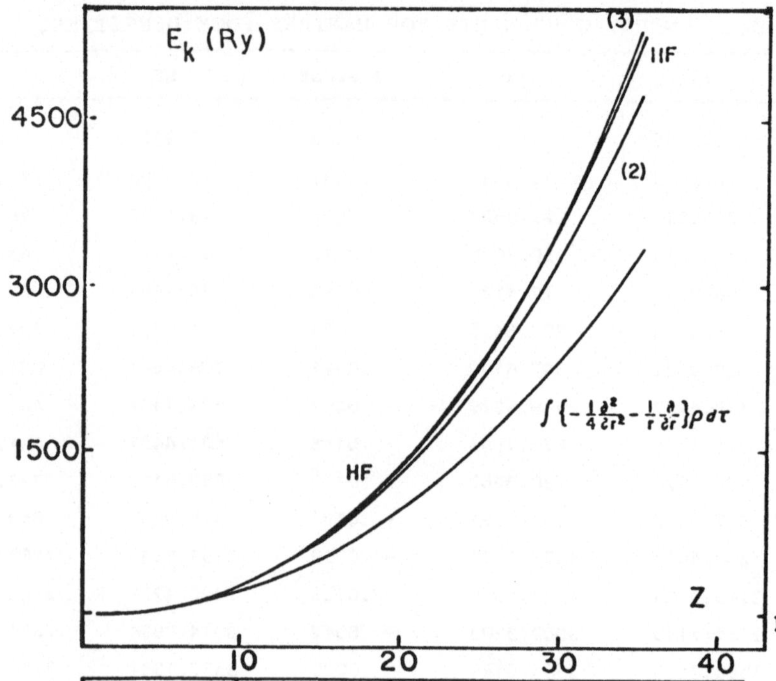

Figure 1

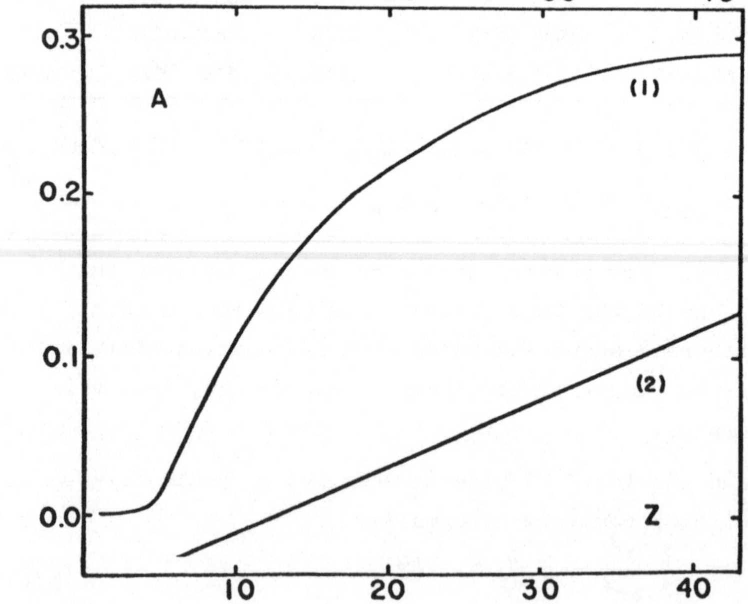

Figure 2.

we obtain

$$\langle \frac{1}{r^2} \rangle = \frac{1.0398}{0.3753} Z^{2.1947} \quad \text{(a.u.)} \tag{83}$$

Finally, Table II shows the values obtained with (79) for the electron-electron Coulomb interaction.

Table II

Electron-Electron Coulomb Energy for Hartree-Fock Atomic Charge Densities Using the $(2/r_{ij})$ Functional and the $dz^{2/3}\rho^{4/3}$ Approximate Functional. d = 1.7921 (Rydberg units).

Z	$1 = \langle 2/r_{ij} \rangle$	$2 = \langle AZ^{2/3}\rho^{4/3} \rangle$	$\frac{2-1}{1}$
2	2.5741	3.1541	0.2253
3	8.2017	7.6045	-0.0728
4	14.4067	13.9227	-0.0336
5	23.2581	22.5667	-0.0297
6	35.7043	34.4717	-0.0345
7	52.3522	49.8783	-0.0473
8	73.2312	69.3661	-0.0528
9	99.6268	93.4348	-0.0622
11	160.1051	152.9021	-0.0450
12	191.7887	185.4778	-0.0329
13	225.8359	221.2618	-0.0203
15	307.1687	304.9702	-0.0072
16	354.1731	353.3905	-0.0022
17	406.2906	406.5086	-0.0005
19	515.4579	523.6388	0.0159
20	570.8364	585.2978	0.0253
23	791.5412	809.5511	0.0228
25	972.4858	988.4735	0.0164
26	1072.396	1089.8655	0.0163
28	1295.166	1304.4992	0.0072
30	1549.740	1549.5837	-0.0001
33	1924.141	1937.3470	0.0069
35	2199.101	2223.4977	0.0111

The results for the exchange energy local density functional have been reported elsewhere [8,9].

References

1. D.A. Kirzhnits, Zh. Eksp. Teor. Fiz. 32, 117 (1957); Sov. Phys. -JETP 5, 64 (1957).
2. E.K.U. Gross and R.M. Dreizler, Phys. Lett. 81A, 447 (1981).
3. J.A. Alonso and L.A. Girifalco, Phys. Rev. B17, 3735 (1978).
4. O. Gunnarson, M. Jonson and B. Lundqvist, Phys. Lett. A59, 177 (1976).
   O. Gunnarson and B.I. Lundqvist, Phys. Rev. B13, 4274 (1979).
5. J. Keller, C. Keller and C. Amador, Lecture Notes in Physics (Springer-Verlag) 142, 364 (1981).
6. J.C. Slater, Quantum Theory of Molecules and Solids, (McGraw-Hill, New York), Vol. 4 (1974).
7. M.S. Gopinathan, M.A. Whitehead and R. Bogdanovic, Phys. Rev. A14, 1 (1976); T.J. Tseng and M.A. Whitehead, Phys. Rev. A24, 16 (1981); T.J. Tseng and M.A. Whitehead, Phys. Rev. A24, 21 (1981).
8. J.L. Gazquez and J. Keller, Phys Rev. A16, 1358 (1977).
9. J. Keller and J.L. Gazquez, Phys. Rev. A20, 1289 (1979); J.L. Gazquez, E. Ortiz and J. Keller, Int. J. Quantum Chem. Symp. 13, 377 (1979).
10. M. Berrondo and O. Goscinski, Chem. Phys. Lett. 62, 31 (1979); M. Berrondo and A. Flores-Riveros, J. Chem. Phys. 72(11), 6299 (1980).
11. R.G. Parr, S.R. Gadre and L.J. Bartolotti, Proc. Natl. Acad. Sci. U.S.A., 76, 2522 (1979); S.R. Gadre, L.J. Bartolotti and N.C. Handy, J. Chem. Phys. 72, 1034 (1980).
12. P. Hohenberg and W. Kohn, Phys. Rev. 136, B864 (1964); W. Kohn and L.J. Sham, Phys. Rev. 140, A1133 (1965).
13. D.R. Murphy and W.P. Wang, J. Chem. Phys. 72, 429 (1980).
14. R.G. Parr and S.R. Gadre, J. Chem. Phys. 72, 3669 (1980).

FREE ENERGY DENSITY FUNCTIONALS FOR NON-UNIFORM CLASSICAL FLUIDS

A. Robledo and C. Varea
División de Estudios de Posgrado, Facultad de Química,
Universidad Nacional Autónoma de México, México 20, D.F.

CONTENTS:

Recent applications of the density functional formalism to the study of the structure and thermodynamics of non-uniform classical fluids are briefly quoted. The problems considered involve fluid-fluid interfaces in pure and multicomponent fluids, fluid to-wall density profiles, solidification, nucleation, spinodal decomposition and interface motions. It is shown that the variational principle for the grand potential yields the potential distribution theory for the equilibrium density, and thus we exhibit the manner in which these two theoretical frameworks to study non-uniform systems are related. Also, we present a derivation of the exact form that the grand potential functional takes for a system of hard rods. Then, we consider attractive interactions in meanfield approximation from which the functional that corresponds to the Van der Waals fluid is obtained.

1. INTRODUCTION.

Even though important advances have accumulated over the last decade which permit a better understanding of the equilibrium properties of uniform classical fluids, the related field pertaining to the structure and thermodynamics of non-uniform fluids, i.e.: fluids for which the singlet density exhibits spatial variation, is relatively less developed. However, there are many physical phenomena of interest that arise from essen-- tial non-uniformities. The following examples of equilibrium situations can be quoted: the structure of fluid-fluid interfaces in pure [1,2] and multicomponent [3,4] fluids, the structure of a fluid in the vicinity of a solid surface [5,6] and the description of liquid cristal phases [7], including a proper cristalline phase[8]. Thus any theory of interfacial tension, of contact angle, wetting and of solidification requires a detailed theory for the statistical mechanics of non-uniform fluids. Amongst important non-equilibrium problems that concern non-uniform fluids is the kinetics of phase separation, [9,10] i.e. the phenomena of nucleation, spinodal decomposition and interface motion are concerned with the time evolution of density fluctuations in inhomogeneous systems.

Most of these applications have been developed along one of two theoretical formalisms. One of these is the density functional formalism [11,12] in which a variational principle for the grand potential determines the number density or singlet distribution function for the equilibrium (or stationary) state. The second approach is that of the potential distribution theory [2,4,12], which provides an expression relating the thermodynamic activity of the system to the equilibrium (or stationary) density. This expression follows from the consideration of a canonical average involving the difference in configurational energy that arises when a molecule is added at $\underline{X}$ to a system of N other molecules. Although it is to be expected that the potential distribution theory expressions must follow from a variational principle for the grand potential, the explicit link between the two formalisms has been provided only until recently. [12]

2. THE FREE ENERGY FUNCTIONAL FOR A NON UNIFORM FLUID.

The most direct approach to the fundamental variational principle for the grand potential Ω for a non-uniform classical system is that recently adopted by Evans[11]. This approach, which is analogous to that employed by Mermin[13] and by Hohenberg and Kohn for the inhomogeneous electron gas, is more naturally expressed in the grand canonical ensemble language. Here, we resume the main argument, but, due to our purposes, choose instead to work with a canonical ensemble.

Thus, we first write the Helmholtz free energy functional as the average

$$F[f_N] = \mathrm{tr}\, f_N\left[H_N + \beta^{-1}\ln f_N\right] \tag{1}$$

where f_N is a probability density in the phase space for a system of N classical particles in a volume V, tr is the classical trace

$$\mathrm{tr} = \left(h^{3N}N!\right)^{-1}\int dp^N dx^N, \tag{2}$$

where p^N and x^N denote momentum and position variables, respectively; $\beta = 1/k_BT$ where k_B is Boltzmann's constant and T the temperature. H_N is the Hamiltonian

$$H_N = K_N + W_N = K_N + U_N + V_N , \tag{3}$$

where K_N is the kinetic energy and W_N the interaction term. W_N is in turn divided into a particle interaction term U_N and an external field term V_N. Eq.(1) merely furnishes the definition of the free energy as an internal energy plus a temperature-entropy term. $F[f_N]$ has the property that for the equilibrium probability density

$$f_N^{eq} = Q_N^{-1} \exp(-\beta H_N) \ , \tag{4}$$

where Q_N is the partition function

$$Q_N = \mathrm{tr} \exp(-\beta H_N) \ , \tag{5}$$

we obtain the usual relation

$$F\left[f_N^{eq}\right] = -\beta^{-1} \ln Q_N \ . \tag{6}$$

Also, $F\left[f_N\right]$ has the minimal property

$$F\left[f_N\right] > F\left[f_N^{eq}\right] \ , \ f_N \neq f_N^{eq}. \tag{7}$$

It can also be proved[11] that for a Hamiltonian with given interactions

$$U_N = U_N(\underset{\sim}{x}_1, \ldots, \underset{\sim}{x}_N) \tag{8a}$$

and

$$V_N = \sum_{i=1}^{N} v(\underset{\sim}{x}_i) \tag{8b}$$

the probability density f_N is a unique functional of the singlet distribution function

$$\rho(\underset{\sim}{x}) = \mathrm{tr}\, f_N \sum_{i=1}^{N} \delta(\underset{\sim}{x} - \underset{\sim}{x}_i) \ , \tag{9}$$

and therefore $F\left[f_N\right]$ is also a unique functional of $\rho(\underset{\sim}{x})$, so that we denote it also by $F\left[\rho(\underset{\sim}{x})\right]$.

In establishing these results[11] it is important to note that, for a fixed interaction U_N, a given probability density f_N is the equilibrium density for an external potential V_N'; i.e. there exists a V_N' such that f_N is the equilibrium distribution for that problem. Thus we write f_N as

$$f_N = \left(\Lambda^N Z_N\right)^{-1} \exp\left[-\beta\left(K_N + U_N + V_N'\right)\right], \tag{10}$$

where $\Lambda = \left[h/2\pi m\beta^{-1}\right]^{3/2}$ is the De Broglie thermal length, and Z_N is the configurational integral

$$Z_N = \int dx^N \exp\left[-\beta\left(U_N + V_N'\right)\right]. \tag{11}$$

Adopting from here on the above expression for f_N (12), the trace in Eq.(1) can be partially performed to yield

$$\mathcal{F}[\rho] = \int d\underset{\sim}{x}\rho(\underset{\sim}{x})\left[v(\underset{\sim}{x}) - v'(\underset{\sim}{x})\right] - \beta^{-1}\ell n \frac{Z_N}{\Lambda^N N!}, \tag{12}$$

and, $\rho(\underset{\sim}{x})$ can be seen to be given by

$$\rho(\underset{\sim}{x}) = N Z_N^{-1} \int dx^{N-1} \exp\left[-\beta\left(U_N + V_N'\right)\right]. \tag{13}$$

It is from Eq.(12) for $\mathcal{F}[\rho]$ that we shall obtain the formula of the potential distribution theory. To this purpose, we consider the variation on the grand potential

$$\Omega = \mathcal{F} - \mu N, \tag{14}$$

that corresponds to adding a particle to a system of other N identical particles, with the constraint that the chemical potential μ is kept constant. This variation is

$$\delta\Omega = \int d\underset{\sim}{x}\delta\rho(\underset{\sim}{x})\left[v(\underset{\sim}{x}) - v'(\underset{\sim}{x})\right] + \beta^{-1}\ln\frac{\Lambda(N+1)Z_N}{Z_{N+1}} - \mu . \tag{15}$$

At equilibrium, and in the thermodynamic limit, $\delta\Omega = 0$ implies the familiar relation

$$\beta\mu_c = \ln\frac{(N+1)Z_N^{eq}}{Z_{N+1}^{eq}} , \tag{16}$$

where μ_c is the configurational chemical potential

$$\mu_c \equiv \mu - \beta^{-1}\ln\Lambda . \tag{17}$$

Now, from Eq. (13) we can rewrite the definition of $\rho^{eq}(\underset{\sim}{x})$ as

$$\rho^{eq}(\underset{\sim}{x}) = \frac{(N+1)Z_N^{eq}}{Z_{N+1}^{eq}} \frac{\int dx^N e^{-\beta\psi(\underset{\sim}{x})} e^{-\beta(U_N+V_N)}}{\int dx^N e^{-\beta(U_N+V_N)}}$$

$$= \frac{(N+1)Z_N^{eq}}{Z_{N+1}^{eq}} \langle e^{-\beta\psi(\underset{\sim}{x})}\rangle_N , \tag{18}$$

where

$$\psi(\underset{\sim}{x}) = W_{N+1}(\underset{\sim}{x}_1,\ldots,\underset{\sim}{x}_N,\underset{\sim}{x}) - W_N(\underset{\sim}{x}_1,\ldots,\underset{\sim}{x}_N) \tag{19}$$

is the difference in potential energy that arises when the (N+1)-th particle is added at $\underset{\sim}{x}$. This, together with Eq. (16), yields the potential distribution formula [2]

$$\rho^{eq}(\underset{\sim}{x}) = e^{\beta\mu_c} \langle e^{-\beta\psi(\underset{\sim}{x})}\rangle_N . \tag{20}$$

This equation relates the equilibrium singlet distribution to the activity $\lambda = \exp \beta\mu_c$, and constitutes a functional relation that determines $\rho^{eq}(\underset{\sim}{x})$. Furthermore, since at equilibrium one has

$$\left.\frac{\delta\Omega}{\delta\rho(\underset{\sim}{x})}\right|_{eq} = \left.\frac{\delta F}{\delta\rho(\underset{\sim}{x})}\right|_{eq} - \mu = 0 , \tag{21}$$

the consideration of kinetic and configurational contributions to $F[\rho]$, i.e.

$$F[\rho] = F_k[\rho] + F_c[\rho] \tag{22}$$

with

$$F_k[\rho] = \beta^{-1} N \ell n \Lambda \tag{23a}$$

and

$$F_c[\rho] = \int d\underset{\sim}{x}\rho(v - v') - \beta^{-1}\ell n \frac{Z_N}{N!} , \tag{23b}$$

leads to the result

$$\left.\frac{\delta F_c}{\delta\rho(\underset{\sim}{x})}\right|_{eq} = \ell n \rho^{eq}(\underset{\sim}{x}) \langle e^{-\beta\psi(\underset{\sim}{x})}\rangle_N^{-1} . \tag{24}$$

Potential distribution theory and the variational principle on the grand potential functional coincide, as they should, in indicating, as seen from Eqs.(20) and (21), that the equilibrium singlet density is that which ensures the uniformity of the chemical potential in the non-uniform fluid. What must be emphasized is that in exhibiting the relationship between the two formalisms we have provided Eq.(21) with an explicit prescription for $\frac{\delta F}{\delta\rho(r)}$ in terms of the interaction potential function.

It is worth noticing, too, that the quantity $\ln \langle \exp -\beta 4 \rangle$ is the classical analogue to the effective potential in the one-electron Schrödinger equation in the Kohn-Sham theory.[14]

3. SOME SPECIFIC EXAMPLES.

We shall now proceed to illustrate how the free energy functional can be constructed from its definition, Eq.(12), for some specific model systems.

A. One-Dimensional Hard-Core Systems.

In order to evaluate the configurational integral Z_N for a systems of hard rods, we look first at its discrete space analogues. Let us consider a linear lattice gas of hard core particles of "length" m, i.e., a particle excludes 2 m + 1 contiguous sites from occupation by other particles. We denote by ρ_s the occupation number or probability to find a particle at s. For the uniform system $\rho_s = \rho = N/M$ for all s, where M is the number of sites in the lattice.

The simplest situation is that of the uniform ideal lattice gas ($m = 0$ and $v_s' = v'$) for which we can write immediately

$$Z_N^{-1}\, e^{-\beta N v'} = \frac{1}{N!}\, \rho^N (1-\rho)^{M-N} = \frac{1}{N!} \left[\rho^\rho (1-\rho)^{1-\rho}\right]^M . \qquad (25)$$

In the equation above $Z_N^{-1} \exp(-\beta N v)$ represents the probability for a configuration of the fluid with uniform occupation number ρ. This probability is equal, after the correction factorial term for indistinguishable particles, to the probability of having the N particles in the lattice, ρ^N, multiplied by the probability of having M-N empty sites, $(1-\rho)^{M-N}$, since multiple occupation is not allowed. If we now let the system to be non-uniform, due to a site dependent external field v_s', the M factors in the last equality above are no longer equal and the probability for a configuration with number density ρ_s is now given by

$$Z_N^{-1}\, e^{-\beta \sum_s \rho_s v_s'} = \frac{1}{N!} \prod_{s=0}^{M-1} \rho_s^{\rho_s} (1-\rho_s)^{1-\rho_s} \qquad (26)$$

When the range of the hard core is extended the occupation of each site is no longer independent from that of its neighbouring sites, but we can still write the probability of an allowed configuration as a site product like in Eq.(26). This product is made by 'building up' the configuration taking as a starting point one end of the lattice and placing the particles in such a way that there are no overlaps of hard cores with those previously placed. For first neighbor exclusion (m=1) we obtain (proceeding from s=M-1 to s=0)

$$Z_N^{-1}\, e^{-\beta \sum_s \rho_s v'_s} = \frac{1}{N!} \prod_{s=0}^{M-1} \frac{\rho_s^{\rho_s}}{(1-\rho_{s-1})^{1-\rho_{s-1}}} \left(1-\rho_{s-1}-\rho_s\right)^{1-\rho_{s-1}-\rho_s}, \quad (27)$$

where the factor $\rho_s^{\rho_s}(1-\rho_{s-1})^{-(1-\rho_{s-1})}$ above is the probability to find a particle at s conditioned to the site s-1 being empty, whereas the second factor, $(1-\rho_{s-1}-\rho_s)^{1-\rho_{s-1}-\rho_s}$ is the probability of finding both sites s and s-1 empty [8]. The other allowed possibility for a configuration, around site s, that of finding site s empty and site s-1 occupied by a particle, is taken into account by the next factor in Eq.(27) (that for site s-1). In general, the result for m-th neighbor exclusion is

$$Z_N^{-1}\, e^{-\beta \sum_s \rho_s v'_s} = \frac{1}{N!} \prod_{s=0}^{M-1} \frac{\rho_s^{\rho_s}}{(1-t_{m-1})^{1-t_{m-1}}} \left(1-t_m\right)^{1-t_m}, \quad (28)$$

where

$$t_m(s) = \sum_{\ell=0}^{m} \rho_{s-\ell}, \quad (29)$$

is the probability of finding the set of m contiguous sites {s,s-1,...,s-m} empty [8]. The free energy functional for this system is therefore given by

$$\beta F[\rho] = \sum_s \left\{ \rho_s \left[\beta \bar{v}_s + \ln \Lambda^{-1}\rho_s\right] + (1-t_m)\ln(1-t_m) - (1-t_{m-1})\ln(1-t_{m-1}) \right\}, \quad (30)$$

whereas the equilibrium density profile is determined from

$$\beta \frac{\delta\Omega}{\delta\rho_s}\bigg|_{eq} = \ell n\rho_s - \beta(\mu_c - v_s) - \ell n \frac{\prod_{k=0}^{m}\left[1 - t_m(s+k)\right]}{\prod_{k=0}^{m-1}\left[1 - t_{m-1}(s+k)\right]} = 0 \qquad (31)$$

Eq. (31) coincides, as it must be, with that derived for this system directly from potential distribution theory [8].

To obtain $F[\rho]$ for a system of hard rods we consider the limiting form of Eq. (30) for large m. Since

$$\frac{1}{N!} \rightarrow \prod_s \left(\frac{e}{N}\right)^{\rho_s} \text{, large } m \qquad (32a)$$

and

$$\left[1 - \frac{\rho_s}{1 - t_m}\right]^{\frac{1 - t_m}{\rho_s}} \rightarrow e, \text{ large } m, \qquad (32b)$$

we have, from Eq. (28) that

$$Z_N^{-1} e^{-\beta\sum_s \rho_s v_s'} = \prod_{s=0}^{M-1} \left[\frac{\rho_s}{N(1 - t_m)}\right]^{\rho_s} \qquad (33)$$

Therefore, for the continuum-space system of hard rods of length σ, we obtain

$$\beta F[\rho] = \int dx\rho(x) \left\{ \ell nA^{-1}\rho(x) - 1 + \beta v(x) - \ell n\left[1 - t(x)\right] \right\} , \qquad (34)$$

where

$$t(x) = \int_{x-\sigma}^{x} dy\rho(y).$$

Functional differentiation of Eq. (34) yields the following relation for the equilibrium density profile

$$\beta \frac{\delta\Omega}{\delta\rho(x)}\bigg|_{eq} = \ell n\rho(x) - \beta\left[\mu_c - v(x)\right] - \ell n\left[1 - t(x)\right] + \int_x^{x+\sigma} dy \, \frac{\rho(x)}{1-t(y)} = 0 \, . \tag{35}$$

Eq.(35) was originally derived by Percus [15] from the grand partition function for this system, and was later obtained [8] from potential distribution theory.

As can be observed, the most relevant properties of the exact free energy functionals derived above are their non-linear and non-local dependence on the singlet density. In contrast with this situation, a truncated gradient expansion, such as that of Cahn-Hilliard, van der Waals yields a functional that, although possibly non-linear, is local in character.

B. Attractive Interactions.

We consider now attractive pair interactions superimposed to the hard-core repulsions, i.e. interactions of the form

$$\psi(|\underset{\sim}{x} - \underset{\sim}{y}|) = \begin{cases} \infty & , \quad |\underset{\sim}{x} - \underset{\sim}{y}| < \sigma \\ \psi_{attr}(|x - y|) & , \quad |x - y| > \sigma . \end{cases} \tag{36}$$

The additional term to the grand potential, Ω_{attr}, and its contributions to the potential distribution formula and to the direct correlation function, that are due to the attractive tail Ψ_{attr}, are easily obtained in mean-field approximation. These are

$$\Omega_{attr} = \int d\underset{\sim}{x}\rho(\underset{\sim}{x})v_{eff}(\underset{\sim}{x})$$

$$= \frac{1}{2}\int d\underset{\sim}{x}\int d\underset{\sim}{x}'\rho(\underset{\sim}{x})\psi_{attr}(|\underset{\sim}{x}-\underset{\sim}{x}'|)\rho(\underset{\sim}{x}') \ , \qquad (37a)$$

$$\frac{\delta\Omega_{attr}}{\delta\rho(\underset{\sim}{y})} = \int d\underset{\sim}{x}\rho(\underset{\sim}{x})\psi_{attr}(|\underset{\sim}{x}-\underset{\sim}{y}|) \qquad (37b)$$

and

$$\frac{\delta^2\Omega_{attr}}{\delta\rho(\underset{\sim}{y})\delta\rho(\underset{\sim}{z})} = \psi_{attr}(|\underset{\sim}{z}-\underset{\sim}{y}|). \qquad (37c)$$

Thus, for the direct correlation function we have the usual mean-field result

$$c_{attr}(|\underset{\sim}{z}-\underset{\sim}{y}|) = -\beta\psi_{attr}(|\underset{\sim}{z}-\underset{\sim}{y}|). \qquad (38)$$

The non-locality of Ω_{attr} is reponsible for the fact that its contribution to c(z,y) is not a delta-like term like in the square-gradient approximation [11]. This is an important difference between the exact (mean-field, in this case) and the approximate van der Waals, Cahn-Hilliard and related theories.

References

1. B. Widom, in Statistical Mechanics and Statistical Methods in Theory and Application V. Landman, ed) Plenum, New York, 1977, p. 33. See also Ref. 10.

2. B. Widom, J. Stat. Phys. 19, 563 (1978).

3. M.M. Telo da Gama and R. Evans, Mol. Phys. 38, 367 (1979).

4. C. Varea, A. Valderrama and A. Robledo, J. Chem. Phys. 73, 6265 (1980).

5. C. Ebner, W.F. Saam and D. Stroud, Phys. Rev. A, 14 , 2264 (1976).

6. D.E. Sullivan, Phys. Rev. B20, 3991 (1979); J. Chem. Phys. 74 , 2604 (1981).

7. I.J. Heilman and E.H. Lieb, J. Stat. Phys. 20, 679 (1979).

8. A. Robledo, J. Chem. Phys. 72 , 1701 (1980).

9. H. Metiu, K. Kithara, and J. Ross, in Fluctuation Phenomena, Studies in Statistical Mechanics VII, (E.W. Montroll and J.L. Lebowitz, eds) North Holland, Amsterdam 1979). p. 229.

10. C. Varea and A. Robledo, J. Chem. Phys. 75, 5080 (1981).

11. R . Evans, Adv. Phys. 28, 143 (1979).

12. A. Robledo and C. Varea, J. Stat. Phys. 26, 513 (1981).

13. N.D. Mermin, Phys. Rev. 137, A 1441 (1965).

14. P. Hohenberg and W. Kohn, Phys. Rev. 136, B864 (1964).

15. J.K. Percus, J. Stat. Phys. 15, 505 (1976).

Lecture Notes in Physics